FOULOIS

ONE MAN AIR FORCE

The Memoirs of

Maj. Gen. Benjamin D. Foulois

FOULOIS
ONE MAN AIR FORCE

The Memoirs of
Maj. Gen. Benjamin D. Foulois

Co-authored by
Col. Carroll V. Glines (USAF Ret.)

NEXCEL Media/Flying Start Group
Dallas, Texas USA

Foulois: One Man Air Force
www.OneManAirForce.com

Acknowledgment is made to the following sources for permission to quote from material already published:

"... and Teach Yourself to Fly." Portions of this chapter originally appeared in *Reader's Digest.*
Reprinted by permission of Keith Elliott.

Library of Congress Catalog Cataloging-in-Publication Data is available upon request.

Table of Contents

Foreword by Roger G. Miller, Ph.D.................................vii

Chapter 1
“ … and Teach Yourself to Fly”.................................1
Chapter 2
My First War...7
Chapter 3
The Philippine Insurrection...................................17
Chapter 4
Lighter-than-Air Days...48
Chapter 5
A Day to Be Remembered.......................................69
Chapter 6
Every Flight a “First”..92
Chapter 7
The Baptism of American Combat Aviation......................111
Chapter 8
With Pershing in Mexico......................................145
Chapter 9
Planning to Fight the First Air War..........................168
Chapter 10
The War Behind the War in the Air............................194
Chapter 11
German Genius and Mitchell Goes Juramentado.................223
Chapter 12
The Time for Patience and Progress...........................239
Chapter 13
The Poverty Days of the Air Service..........................266
Chapter 14
The Truth about the Air-mail “Fiasco”........................282
Chapter 15
The End of the Beginning.....................................314

Epilogue...355

Books by Carroll V. Glines

The Doolittle Raid (Schiffer) *
Round-the World Flights (1st Edition 1982 (Van Nostrand Reinhold)
2nd Edition (1990) (TAB Books)
3rd Edition (2002) (Potomac Books)
From the Wright Brothers to the Astronauts (Gen. B. D. Foulois autobiography) (McGraw-Hill)
Jimmy Doolittle: Daredevil Aviator and Scientist (Macmillan)
The Compact History of the U.S. Air Force (Hawthorn)
The Modern U. S. Air Force (Van Nostrand Reinhold)
Doolittle's Tokyo Raiders (Van Nostrand Reinhold) *
Four Came Home: Doolittle's Lost Crews (Pictorial Histories Publishing Co.)
Air Rescue! (Ace Publishing Co.)
Jimmy Doolittle: Master of the Calculated Risk (Van Nostrand Reinhold)
The First Flight Around the World (Pictorial Histories Publishing Co.)
Grand Old Lady: The Story of the DC-3 (Pennington Press)
Our Family Affairs (Walker Press)
The Complete Guide for the Serviceman's Wife (Houghton Mifflin)
Helicopter Rescues (Scholastic Press)
Lighter-Than-Air Flight (Watts)
Polar Aviation (Watts)
Minutemen of the Air (Random House)
The Legendary DC-3 (Bantam)
The First Book of the Moon (Watts) *
The Wright Brothers: Pioneers of Power Flight (Watts)
Attack on Yamamoto (Schiffer) *
The DC-3: The Story of a Fabulous Airplane (Lippincott)
The Saga of the Air Mail (Van Nostrand Reinhold)
I Could Never Be So Lucky Again (Doolittle autobiography) (Schiffer)
The Evergreen Story: History of the World's Foremost Aviation Enterprise (Library of Congress)
Flights: Aerospace Industries and American Well-Being (Community Communications)
Roscoe Turner: Aviation's Master Showman (Smithsonian Press)
Chennault's Forgotten Warriors (Schiffer)
The Amazing Gooney Bird (Schiffer)
Bernt Balchen: Polar Aviator (Smithsonian Press)
Around the World in 175 Days (Smithsonian Press)
Gen. James H. Doolittle: Master of the Calculated Risk (Pictorial Histories)
Those Legendary Piper Cubs (Schiffer)

* Also published in foreign languages

Foreword

B. D. F.
The Last of the First

It is an honor to be asked to help introduce a new generation of readers to Maj. Gen. Benjamin Delahauf Foulois, a pioneer aviator who learned to fly at the inception of flight, a senior leader of the U.S. Army Air Service and Air Corps, and a founding father of the military airpower embodied in today's U.S. Air Force.

I have always thought well of Benny Foulois. (The proper pronunciation of his French last name, by the way, is "fool-wah", but he pronounced it "fool-oy.") He was a tough, unpretentious, practical-minded former enlisted man whose spirit of adventure, ability to work with his hands, and interest in technical subjects led him first into the U.S. Army and then into flying at a time when military aviation was just getting off the ground—so to speak. Foulois combined dedication, initiative, and energy with formidable ambition. He earned an officer's commission during the fighting in the Philippine Islands at the turn of the twentieth century, and through courage, intelligence, and hard work—and a healthy dose of good luck—rose to the rank of major general and to command of the U.S. Army Air Corps. Foulois presents his Horatio Alger story in this autobiography.

Can't remember now when I first read Foulois's autobiography. The U.S. Air Force commissioned me in 1969—the year after Colonel C. V. Glines first published the book entitled *From the Wright Brothers to the Astronauts*—and I already had begun to take a serious interest in the history of my chosen service, so it might have been right then. I certainly was familiar with the autobiography by 1980 when I became a civilian historian in the U.S. Air Force History Program and knew it intimately by 1993 when I joined the Air Force Historical Studies Office, the book-writing element of our program. Still later, the autobiography served me as an indispensable reference during the commemoration of the Centennial of Flight from 2000 to 2003, and it has prominently contributed to many of the books, monographs, articles, and essays I have had the opportunity to write over the years. The autobiography is not without its idiosyncrasies, however, and some of these are worth pointing out:

First and most important, Foulois's arch rival throughout his aviation career was the great prophet of independent air power, William "Billy" Mitchell. The two men were almost exact opposites. Billy was the articulate, dynamic, charismatic scion of a wealthy, prominent family; Benny, the down-to-earth, blue collar son of an immigrant plumber with few social graces and fewer pretensions to high society. He preferred whiskey to cocktails and poker to polo, and often was as much a mechanic and technician as pilot. The two men shared an unshakeable belief in the future of airpower and deep commitment to an independent air force that should have made them allies, but profound personal ambition ensured their enmity. To say they disliked each other is to exercise considerable understatement.

Mitchell's fame eclipsed that of Foulois, and Benny considered that an injustice. He saw Billy as a "zealous and jealous man by nature" who had "attacked me viciously many times in print," and he used his autobiography to redress the balance. "It is my intention to set the record straight more than once in regard to Billy Mitchell in these pages," he promised, and "[f]or the first time I am going to tell my side of the story in the interest of historical accuracy." (p. 46) Ultimately, Foulois surpassed Mitchell career-wise. Benny commanded the U.S. Army Air Corps, while Billy never rose beyond deputy, and Foulois retired as a major general while Mitchell left the service a colonel. Little matter. Foulois insisted upon having the last word, and his dislike of Mitchell permeates the autobiography.

Of interest to the historian, Foulois's autobiography almost completely ignores his personal life outside his military experiences. Reading the book, for example, one would almost never know that he married—twice. Through his first wife, Ella

Van Horn, Foulois united with a distinguished military family. But that marriage ended in divorce and many in the "Old Army" came to believe that Foulois had treated Ella badly, a belief that damaged his reputation and standing in the officer corps. Foulois remained a bachelor for many years afterward, but in 1924 married a lady from Philadelphia, Elizabeth Grant Bethel, who has been described as both "formidable" and "determined." Benny and Elizabeth remained a devoted couple until her death.

A great strength of the autobiography lies in that Benny Foulois was a resourceful raconteur, a wonderful story teller, and his colorful descriptions of the events he experienced, the places he went, and the individuals he met are a major attraction of the narrative. But like many old veterans—and some not so old—his stories got better with age. It is well to be aware that many of Foulois's accounts bear the hallmark of careful polishing through frequent repetition.

In his introduction to the original autobiography, distinguished cartoonist Milt Caniff, creator of "Steve Canyon" and "Terry and the Pirates," referred to Foulois as "the Little General" and that was an apt description. He was a small man, a fact that at least once worked in his favor. In 1909, Orville Wright wrote Wilbur from Ft. Myer, Virginia, where the Wright Military Flyer was undergoing acceptance tests, that he liked the lieutenant very much. "He is a little fellow only weighing 130 lbs," Orville reported to his brother, and he invited Foulois to be his passenger on the cross county flight that the Flyer had to perform to meet the requirements of the U.S. Army contract. The minimum speed required was 40 m.p.h. with two men, and the Wrights would receive an additional $2,500 for each m.p.h. attained over the minimum. The Flyer averaged 42.5 m.p.h. during the test, earning them $30,000, $5,000 above the contract price of $25,000—a huge amount in those days. Had Orville's passenger been a larger, heavier man it is doubtful they would have earned the extra money; quite possibly, the minimum speed might not have been achieved at all. But thanks to Benny, the Wrights succeeded. It would be nice to report that the Brothers awarded the lieutenant a share of the earnings, but they were not that generous.

We read history and documents such as personal memoirs to learn about our past; to become familiar with our roots; to gain some understanding of where we came from and how we got where we are today. In this perspective, Foulois's autobiography is significant because, while the early twentieth century was a time much different from our own, in many ways the experiences, challenges, and controversies over preparedness for defense and the application of military airpower seem timeless. Then, as now, technology had compressed time and distance, and efforts to identify, adopt, and apply new technologies led to a host of controversies over roles and missions, organization and functions, and funding priorities. Personalities and political leadership proved critical factors in determining what can be accomplished and what is left undone. In the 1920s and much of the 1930s, the Air Corps lacked the direct threat of an obvious, credible enemy that justified independence, a situation that exists to some extent in this post-Cold War era. And at the beginning of the 1930s the Great Depression devastated the national economy; military preparedness took a back seat to putting the country to work; Congress and successive Administrations pinched pennies; and airmen had to do without the modern technology they considered necessary to achieve their airpower mission. Parallels with recent trends and events seem obvious. One must conclude that Foulois's autobiography speaks to today's airpower leaders.

Benny Foulois lived a long, adventurous and useful life. In 1935, he and Elizabeth retired to Ventnor, New Jersey, and his retirement was sufficient that he did not have to earn a living. He, thus, could devote his time and energy to speaking to civic, business, and educational groups on the importance of airpower and military preparedness. In 1941, Foulois campaigned unsuccessfully for a seat in the House of Representatives, and in 1942 he turned down an invitation to return to active duty. Instead, Foulois assisted Mayor Fiorello LaGuardia of New York City establish a civil defense program and helped organize the Civil Air Patrol, which remains an important national organization today. After World War II, Foulois continued to speak and write on behalf of airpower and national defense. For many years he served as president of the Air Force Historical Foundation. In 1959, Gen. Thomas D. White, Chief of Staff of the U.S. Air Force, invited Foulois to reside on Andrews AFB outside Washington, D.C. Benny Foulois spent his last years living on Andrews, garnished with honors, traveling and speaking to U.S. Air Force audiences throughout the world.

Benjamin D. Foulois was a hero of early military aviation, and he tells his story well. His autobiography is crammed with interesting events, exciting adventures, and colorful stories that explain, illuminate, and illustrate the dawn of flight and in the origins and development of modern airpower. Foulois presents an important account in an entertaining manner. His autobiography is a "must" for anyone interested in the history of military aviation. I urge you to read and enjoy!

Roger G. Miller, Ph.D.
Air Force Historical Studies Office
Washington, D.C.

Benjamin Delahauf Foulois

Chapter One

"... and Teach Yourself to Fly"

"Uncle Joe" Cannon, Speaker of the House of Representatives, was a skeptical man. On a warm and windless July day in 1909 he saw his first airplane at Fort Myer, Virginia. He muttered to a bystander, "You can't convince me that that thing will fly." Moments later the plane was catapulted into the air. "Well, it's flying," he admitted, "but you can't make me believe it will stay up!"

Nevertheless, it did. It flew serenely to Alexandria, turned around, and returned to Fort Myer. History does not record Uncle Joe Cannon's reaction to its safe, uneventful landing.

Two men climbed off the airplane (it had no cockpit), and elatedly compared notes. I was one of them. The other was the plane's pilot, a partner in its design and manufacture, and his purpose in making the flight was to sell his airplane to the United States Army. His name was Orville Wright.

Proudly, we shook hands. In a single flight we had established three world records: for speed (42.5 miles per hour), distance (10 miles), and altitude (400 feet).

My report, as official Army observer on the test flight, satisfied my superiors as to the plane's performance. And since the Wright brothers had already met the first specification—that the airplane be small and light enough to be transported by horse and wagon—the Army paid them $25,000 for their record-breaking craft, plus a $5,000 bonus. Now all the new air service needed was a pilot. As it turned out, I was to be it.

I was not entirely unqualified. Although the speed and cross-country final acceptance test on July 30, 1909, was my first flight in an airplane, I later logged fifty-four minutes of training time with Wilbur Wright after the final test. He even allowed me to handle the controls. But I had never been up in the machine alone, taken it off, or landed it.

"Don't worry," said Gen. James Allen, chief Army signal officer, "you'll learn those techniques as you go along. I've decided to send you and the flying machine to Texas. The weather is better there."

"Very good, sir. Have you any special instructions?"

"Your orders are simple, Lieutenant. You are to evaluate the airplane. Just take plenty of spare parts—and teach yourself to fly."

Before leaving for Texas, I received a voucher for funds to buy parts. It was the nation's first air force budget, and it was earmarked for one year's maintenance. The amount: $150.

When I stepped off the train in San Antonio on the morning of February 5, 1910, I fear I did not cut too glamorous a figure as sole custodian of United States aerial might. At thirty, I was several years older than the average student pilot of today. Moreover, as the *San Antonio Express* observed, "Lieutenant Foulois is one of the smallest men in the Army. He is five feet five inches in height, and he weighs only one hundred thirty-five pounds."

Much of my time at San Antonio's storied Fort Sam Houston that spring was spent writing to Orville Wright, asking him how to execute basic maneuvers, how to avoid basic disasters—in short, how to fly an airplane. As far as I know, I am the only pilot in history who learned to fly by correspondence.

Like the plane the Wrights had first flown at Kitty Hawk, Army *Aeroplane No. 1* was the "pusher" type. Its two chain-driven propellers, located behind the pilot, were powered by a four-cylinder engine developing, under ideal conditions, a thundering 25 horsepower. A historian of that epoch wrote that the world's first military airplane looked like "a collection of bamboo poles, more or less indefinitely attached to a gasoline engine."

Nevertheless, this airplane had already killed one man. Lieut. Thomas E. Selfridge was the first man to die in a plane crash. The smashup, during the Fort Myer tests, had also severely injured Orville Wright. But for this accident, my friend Tom would probably have been the first military aviator to fly a military airplane. Now, in the same machine, modified and rebuilt, I was to have the distinction.

Old *No. 1* came by express to San Antonio in seventeen wooden crates. It was painstakingly reassembled by eight enlisted men under my command. We worked on our craft in a shed on the cavalry parade ground to the gibes of hard-bitten cavalrymen. ("Whatcha making, boys—a kite?") They were the Army's elite. We were referred to as its "crazy birdmen."

But the old-timers at Fort Sam indulged our madness, even aided us as we improvised our way into the wild blue yonder. A blacksmith lent his forge and skill to help us fabricate parts. A tailor helped us sew linen for wing, elevator, and rudder surfaces. A plumber secured pipe fittings to fuel lines.

Target date for my first flight was March 1. That morning we pushed *Aeroplane No. 1* to a single track fronting its "launching tower," a 20-foot steel structure with a 1,400-pound weight rigged to a pulley on top. The pulley system was connected to the airplane in such a way that when I pulled a trigger the weight would fall, snapping me and my 800-pound flying machine along the monorail, and, with luck, aloft.

But the launching weight was too much for the pulley. A bolt broke, and history was deferred by a day while our blacksmith forged us a stronger one.

A soft breeze blew in from the southwest at eight miles an hour on Wednesday, March 2. By 9:30 A.M., an audience of about 200, counting hooting cavalrymen, had assembled on the edges of the parade ground, discreetly keeping its distance from the strange contraption on which I perched. Sgt. Stephen J. Idzorek and Corp. Vernon L. Burge stationed themselves at the propeller and cranked the engine, hacking and sputtering, to life.

Eight men delicately lifted *Aeroplane No. 1* and Army Aviator No. 1 to the launching track. Two men remained, one at either wing, to keep the plane balanced precariously on the track.

I couldn't discern it above the noise of my engine, but reporters said "a reverent hush fell over the crowd." This was appropriate, for I was praying.

I extended my arms winglike to tell the crew that the machine was in balance and I was ready for takeoff. Burge gave me a "thumbs-up" sign for luck. I pulled the trigger to release the weight on the tower, and the plane lurched forward.

We roared along the 50-foot length of monorail at tremendous speed—possibly as much as 30 miles an hour. Gently, following Wright's instructions, I tugged on the stick controlling the elevator surfaces. Then, suddenly, I was air-borne—aloft alone for the first time.

My first flight of a government-owned plane took an eternity, lasting just seven and a half minutes. Soaring above soldiers, tents, horses, I discovered my natural element, the sky. But my exhilaration was offset by misgivings at the thought of landing. "Coming down," Wright had warned me, "is the most critical part of flying."

While townspeople waved and shouted below, I circled inside the parade ground—perhaps two miles square—six times, rising once to an altitude of 150 feet. ("Was it difficult breathing up there?" I was asked later.)

But I could not postpone landing indefinitely. At the end of my sixth run I descended toward our shed. Landing had to be accomplished with the engine off, or "dead stick," and my hand was on the gas cutoff as I lost altitude. Suddenly, just as I killed the power, an automobile lumbered into my path. It was driven by a kindly old doctor, arriving late for the show.

"With the coolness of steel," a newspaper reported, "Lieutenant Foulois managed to avoid contact with the automobile." Actually, all I did was to jerk instinctively on the control stick, and we literally hopped the good doctor's flivver with the last of my flying speed.

A moment later I was down a half mile beyond my planned point of touchdown. Friends, curiosity seekers, even cavalrymen, rushed onto the field to congratulate me. Inasmuch as I could walk away from it, I could not help but agree with them that I had executed a "splendid landing."

Somewhat astonished by my success, I made three other flights that day. The last one ended disastrously. The fuel line broke, the engine stopped, and old *No. 1* hit the dust in a jarring landing, shearing her rudder and shaking up her novice pilot. I had executed my first takeoff, my first solo, my first landing, and my first crash on the same day.

In the months that followed, Army *Aeroplane No. 1* became a familiar, grotesque figure in the San Antonio sky. The light craft often bucked like a mustang in the unpredictable South Texas thermal currents, so the saddle maker of a field artillery battery modified a trunk strap to hold me to my seat. It worked perfectly. We had devised a safety belt.

Application of wheels to aircraft was another innovation for us. The biplane couldn't take off in a cross-wind on its ski-type skids; one effort to do so had put *No. 1* in the shop for a week. So we decided to experiment with wheels under the plane. A set from a farm cultivator worked splendidly, and thereafter wheels became standard equipment.

I crashed frequently. Yet in only one accident, when a truss wire pinned my leg to the ground, was I injured. The resultant slight scar is the only mark I bear today from twenty-eight years of flying. But old *No. 1*'s injuries were more serious. So was the plight of the nation's air force budget: my $150 maintenance fund dwindled quickly. I was forced to dip into my own pocket to purchase material and parts for new wings, new rudders, a new elevator, a new propeller. All told, I spent $300 of my personal funds keeping the one-plane, one-pilot air arm flying.

Then came a day in April 1911, when tired old *No. 1* was replaced by a sleek new Wright Model B Aeroplane. A famous exhibition pilot, Phil O. Parmelee, came with it. At last I was to receive formal instruction in the art of flying. I had made sixty-four flights, logging nine hours and ten minutes of flying time, when I took my first lesson under a "professional."

The Wright brothers eventually restored Army *Aeroplane No. 1* and presented it to the Smithsonian Institution at Washington, D.C. It hangs there today, alongside the Wrights' original Kitty Hawk. Since those primitive days, when every flight was an experiment and an aviation "first" of some kind, I have watched every military aeronautical development with great personal interest. I made, and participated in the making of, decisions which have shaped not only the Air Force we have today, but general aviation as well. I made mistakes—lots of them. But I learned from them. I made some correct judgments, and I'm glad I made them. I was demoted for making some, too. Again, I learned.

What disturbs me today, and what prompts me to write my memoirs as I near the end of my ninth decade, is that I see too many unfavorable developments in our international relations, our government, and our military services that are identical to what I saw in the Spanish-American War, the Mexican Punitive Expedition, World War I, World War II, and the Korean War. Our nation and others are again involved in a war in Asia, and again an enemy lurks dangerously close to our back door. We are involved militarily in dozens of trouble spots around the world. We have more than 3 million men under arms today. We live in an armed society, but in spite of experience in warfare, we are still making mistakes—the same mistakes we made in other times and in other wars.

In the pages that follow I will tell of the successes and the mistakes we made during my tenure as a military aviator, the lessons learned, and my observations about today and tomorrow.

Chapter Two

My First War

As I look back over almost nine decades and marvel at how I lasted this long, I realize that it was my parents who had the greatest influence on me in my growing-up years. As every boy should, I learned the difference between right and wrong from them, and I owe whatever success I've had in later life to their homely teachings. It was their practical approach to solving life's problems that I inherited, along with their energy and love of life.

My father was born in Paris in 1850 and served in the Franco-Prussian War of 1870-1871. He had been a tinsmith before his military service and later turned to plumbing when it looked as though cities might use metal pipes to solve their sewage and water problems. But the hardships of war disenchanted him with the life of a Frenchman and he migrated to America as soon as he was discharged. He located in Pittsfield, Massachusetts, then Washington, Connecticut, where he started his own plumbing business. He met my mother, a Boston-born nurse, and they were married in 1876.

The first child of this marriage was a boy born a year later, named William Henry after some favorite relative on my mother's side. A girl, Florence Augusta, came the next year, and I was born a year later on December 9, 1879. It was my father's turn to remember his relatives, and I was named Benjamin Delahauf after a favorite uncle.

Another boy was born in 1881 and was named Burchaell, again after a relative on my mother's side. He was not so healthy as the other three of us, and I remember little about him except that he never played with us. He died at age sixteen of spinal meningitis.

Except for Burchaell's poor health, I suppose the Foulois children had a normal life for New England children growing up in the 1880s. We all attended the same one-room schoolhouse from the time we were five years old. It was as typical a schoolhouse as any of the period. The same teacher, Randall Davidson, taught us year after year, and managed to widen our horizons as best he could when he had two dozen students, all of different age levels and talents. A stern disciplinarian, his favorite instrument for child alignment was an eighteen-inch ruler. He used it as a pointer, bookmark, stove stoker, attention getter, and, most important, as a disciplinary tool which was applied to our posteriors with great force and accompanied always by a lecture for the benefit of all observers.

I am not ashamed of the fact that I spent more time outdoors than I ever spent under a roof during my youth. I couldn't wait to feel the breeze against my cheek after having been cooped up inside the schoolhouse. I refused to take my lunch to school because I wanted to run home, about a mile distant, just for the exercise and the sheer fun of running.

Fishing, sports of all kinds, especially baseball, and bike riding were my great loves when it came to exercise, but I guess my greatest personal fun came out of just being in the woods, free from care and worry. I marveled at the animals I stalked—how they lived in spite of man and his ever-encroaching ways on their freedom of movement. Birds fascinated me most of all, and I would lie in the woods on summer afternoons by the hour and watch them work and play.

"Why can't man fly?" I asked myself over and over. "Birds aren't very smart, yet they can fly. People are supposed to be smarter than birds, and yet we can't get ourselves off the ground. Maybe someday people will learn to fly, and what a great day that will be!"

Little did I *really* think that men would be able to go from one place to another in the air, although I had heard about balloons and things called dirigibles but had never seen them. And everything I had heard about them was bad.

It was a foregone conclusion that the boys in the family would one day enter Dad's business. My father was the only plumber within many miles. His principal job seemed to be the installation of sewer pipes and septic tanks, but he had dreams of bringing an improved water system to the town of Washington. He thought of piping water from streams and springs just as they did in the big cities so that our town wouldn't have to use wells or have the water carted from nearby springs.

Father's business prospered in the early 1890s—so much so that my brother William quit school to learn the trade. I stayed in the one-room school for ten years, and then attended Gunnery Preparatory School for one year, but the fever to go to work in my father's shop soon infected me, too.

"Son," my father said to me at the end of my eleventh year of formal schooling, "you will soon have to make up your mind whether you want to make a living with your head or your hands. If you want to use your hands as I do, you need no further schooling. If you want to use your head, you'll have to go on with more book learning. The choice is yours."

The choice wasn't hard for me. I liked books, but the only men I knew who used their heads instead of their hands were the schoolteacher and the preacher. They didn't make much money, and Dad's business was thriving. I decided to be a plumber.

My apprenticeship was thorough. My father insisted that I must learn to do everything in the shop and do it right. "You owe it to the family name to do a job that you and your children will be proud of," he used to say.

I learned to wipe a joint, cut and thread pipe, lay sewers, and caulk joints. I learned how to estimate vertical, horizontal, and angular measurements by eye, move heavy objects by using leverage, and use all kinds of hand tools. In short, I learned how to use my eyes, hands, and feet to maximum advantage. The skills I gained gave me confidence. The coordination of mind and body, the joy of physical labor, the satisfaction of doing a good job, and the knowledge that I could earn a living on my own, all contributed to my mental well-being.

After about a year of working with my father, I began to feel a restlessness that every boy experiences sooner or later. That restlessness really began when I was selected, along with another teenager, by the Christian Templars Club to attend a convention of the organization in New Haven.

The trip showed me that there was a world outside of the familiar town of Washington—a world that I wanted to see. Consequently, when I received an invitation to visit New York in 1897, I accepted.

The invitation had been extended by two girls I had met who lived and worked in New York but came to our town for summer vacations. I had dated both of them and they promised that if I could come to New York that spring they would give me a guided tour. They suggested that I could stay at the Sailor's Mission while I was there for fifty cents a day.

One of them had a sailor boy friend and the other had a brother in the Navy. Both girls had been thoroughly brainwashed with sailor talk and they gave me my first taste of Navy "public relations." I don't know how the Navy does it, but they are perfectionists at the propaganda art—so good, in fact, that no one in my part of the country had ever heard of the peacetime Army in the 1880s and nineties. But these girls knew all about the Navy and talked Navy the whole time I was there. They infected me with their talk so much that I went back home with "Navy" on my mind.

I was just becoming aware of national events in my late teens and knew that the United States in the fall of 1897 had worked itself into a corner diplomatically. Two years before, Richard Olney, Secretary of State under President Cleveland, had asserted the Monroe Doctrine in regard to a dispute between England and Venezuela over the boundary of British Guiana. The same year an insurrection broke out in Cuba, and Americans sided with the insurgents rather than the Spanish. The recriminations by the Spanish under Gen. Valeriano Weyler led to more sympathy for the insurgent cause. Cleveland did nothing to intervene during his administration. McKinley, his successor, had no firm policy and was content to let matters drift.

It remained for the Spanish Minister to provoke the people of the United States. Dupuy de Lome, Spanish Minister to the United States, had written a letter in which he characterized McKinley as "a weakling... a bidder for the admiration of the crowd," and "a would-be politician who tries to leave a door open behind him while keeping on good terms with the jingoes of his own party." This letter was intercepted and given wide publicity. The newspapers said the nation's President had been insulted; the Spanish should pay.

A second incident that received the most public notice was the mysterious sinking of the battleship *Maine* in Havana Harbor, on February 15, 1898. An explosion ripped the vessel from stem to stern, and killed 260 members of her crew. Spanish treachery now had to be avenged.

It was the sinking of the *Maine* that caused me to become really aware of current world events for the first time in my life. Here was a war coming in my lifetime. Maybe I could fight as my father had done in Europe and the men in his shop had done in the War Between the States, I thought. The Navy needed men. This was my chance.

The next month after the *Maine* had been sunk, I prevailed upon my parents to let me make another trip to New York. My railroad friends let me ride their engines to New Haven and then to New York. I stayed at the Sailor's Mission again and learned that the Navy was recruiting on board the receiving ship *Vermont* tied up at the Brooklyn Navy Yard. I hastened to the ship.

"Can you box the compass?" I was asked by a burly Navy chief.

I didn't know what a compass was and wondered if he wanted me to construct a box or crate for something.

"I don't know what you mean," I replied.

"Can you heave the lead?"

I had no idea what he was talking about and shrugged.

"We don't want any landlubbers, little mate," I was told. "Besides, if you're under twenty-one, you have to have your parents' permission to enlist."

I was keenly disappointed.

An ultimatum had been sent to Spain on March 27 which demanded the right of the United States to act as arbitrator between the Spanish government and the insurgents. On April 19 a resolution was passed by the Congress, proclaiming Cuba free and independent and authorized the President to use land and naval forces to expel the Spanish from the island. The President signed it the next day which made it a declaration of war. Two days later the Navy was off to Cuban waters to blockade the island. On the same day the Congress voted to double the size of the regular Army, and issued a call for 125,000 volunteers from the several states. More Navy volunteers were needed.

This news made me all the more restless, and I was sure the Navy would take me now. On a quiet Sunday morning in May, I slipped my brother's birth certificate out of the family Bible, packed my bicycle saddlebag, and left. I rode all day, and arrived at Jersey City where I looked up an old sailor I had met on the New York water front during my previous visit. He lived in a shack along the river, and let me stay with him. I hiked to the Brooklyn Navy Yard next day, and was turned down as before. Whatever the Navy meant by "boxing the compass" and "heaving the lead," they must have been important.

I tried to find a job, but no one would have me. By accident, I passed an Army recruiting station where volunteers were being sought for the 1st United States Volunteer Engineers. The brain-washing by my two Navy-recruiting girl friends had been so complete that joining the Army had never once entered my head. I was curious as to what the Army was like and what it did. I went inside to enquire, and fifteen minutes later, much to my surprise, I was in it.

What had sold the Army on me was that I was a plumber and could presumably use my hands. The Corps of Engineers needed men who could build bridges, roads, privies, or whatever. Plumbers might be useful, or at least could be trained to be so.

What clinched it was my brother's birth certificate. When the recruiting sergeant asked me how old I was, I just pulled out my brother's birth certificate without saying a word. I just couldn't tell a lie so I let the paper speak for me. As a result, I was enlisted under the name of William Henry Foulois, and my age was listed as twenty-one. I had to stretch a little to pass the height requirements, but I made it.

I was kept around the recruiting office for several days and helped with clerical duties while other volunteers were recruited. When asked what I liked to be called, I always answered “Ben,” and never made any effort to hide my identity. It was the sergeant's error and not mine, I reasoned. I didn't learn what a fraudulent enlistment was until several years later.

When a large enough group was recruited, we were sent to Camp Townsend near Peekskill, New York. Shortly after arrival I was made a lance corporal and given my first real responsibilities.

Training for the next month was intensive. We learned how to build bridges, revetments, roads, and fortifications. Everything was what we would call on-the-job training today, where we learned the skills we needed by doing. There was no classroom instruction. Regular Army sergeants told us and showed us. Since all of us were men who had used our hands before—carpenters, ship fitters, steeplejacks, masons, and plumbers, we didn't need much instruction after once being shown. I learned how to use a transit rod and level with the surveying section since there was very little plumbing work to do. This training was to come in handy many times in the years to come.

In early July 1898, we sailed from New York for Puerto Rico aboard the *City of Chester,* an old tub that had been pulled out of storage and caulked up hurriedly. The heat below decks, where the enlisted men were quartered, was fierce. “Only horses and officers on deck,” we were told, which enabled volunteer enlisted men to see where they ranked in the scheme of things.

While we were en route to Puerto Rico, the famous battle of San Juan Hill had been fought, and the American fleet had overwhelmed the Spanish fleet in Santiago Harbor. The Americans had one man killed and one seriously wounded. The Spanish had 323 men killed and 151 wounded. More than 1,800 prisoners had been taken: Adm. William T. Sampson, elated with these statistics, cabled Washington:

> "The fleet under my command offers the nation, as a Fourth-of-July present, the whole of Cervera's fleet."

This was a splendid example of Navy publicity. Typically, the Army was having a rough go of it on land as Maj. Gen. William R. Shafter's troops tried to maintain a precarious toehold in the city of Santiago. By the time we had reached Ponce, Santiago had finally been captured.

The expedition to Puerto Rico was supposed to be a major campaign for us, but the surrender of Santiago and the loss of their fleet had dampened Spanish enthusiasm for more war. On July 25 the first American troops started to land at Guanica, a small village near Ponce. My unit followed several days later, and our infantry took formal possession of the city of Ponce as the Spaniards withdrew toward San Juan on the other side of the Island. Our job as engineers was to rebuild damaged bridges, lay out fortified camps, and do whatever was necessary to get our troops settled so they would have a base of operations from which to begin the campaign against the enemy.

As the history books record, there were no great battles in Puerto Rico. The most serious of the few skirmishes took place at Coamo where the Spanish lost six men and the Americans none. By August 12 the Spaniards had had enough war, and word was received that the armistice had been signed.

The war was not over by any means, as far as we engineers were concerned. The real enemy in Puerto Rico was disease. Malaria, yellow fever, typhus typhoid, dysentery, and diarrhea disabled our forces just as effectively as enemy bullets—more so, because it took many more men to care for the sick than would have been necessary if they had been wounded on the battlefield.

It was unnerving to see healthy, enthusiastic men drop and be carried away from the camp feet-first. It was here that I remembered what my mother had told me about quinine and tropical diseases. She had once admonished me, "If you ever go to the tropics, take plenty of quinine with you if you never take another thing." I remembered this, bought some just before we had sailed from New York, and took daily doses the whole time I was in Puerto Rico. It kept me alive and on my feet.

We began our reconstruction program of road and bridge building as best we could with an ever-decreasing work force. The later arrival of quinine helped save many. Although we were miserable from the heat, we were told that Puerto Rican rum could make us forget our misery at night. However, at day's end it was the infantrymen who would stay up late swigging the stuff. Engineers were too tired even to tip up a bottle.

During the building program, Capt. Samuel Rodman, our commanding officer detached me in charge of 20 men to the village of Yauco at the request of the mayor there. Puerto Rican bandits had been robbing the farmers heavily, and the mayor feared an assault on his town any day. He had asked for protection and Rodman was ordered to give it.

My men and I were issued new Krag-Jörgensen rifles before leaving. To us this was a mark of prestige since only regulars were given these beautiful rifles, never volunteers. Most of us had never fired a rifle before, but it made us feel good to carry one, anyhow.

We proceeded by railroad and then pack train and arrived in Yauco where we found that the mayor was a big wheeler dealer. He owned most of the town and surrounding farms. He had two beautiful daughters and promptly offered either one to me if I would protect his town. Duty meant more to me than daughter so I respectfully declined. I set up guard posts around the town and organized patrols in the countryside. We captured several bandits and turned these over to the local police, who promptly shot them.

My service lasted just seven months in Puerto Rico. The rum and the quinine, self-discipline, and luck kept me alive. Of the 40 non-coms in my regiment, only my buddy, Charlie Towle, and I survived to sail home in January, 1899.

Taking sixty grains of quinine per day took its toll on my nerves. By the time I reached New York, I was a shaky wreck. Once I passed the health officials and was let ashore, I went straight home on leave to our family doctor in Connecticut. He prescribed two things for me: beer and cold baths. I didn't like beer, so my family drank the beer and I took the cold baths. In a few days I felt better, and the doctor took great pride in his prescription for my recovery.

After a short leave I returned to New York and was mustered out of the service. I was so sick of Army life that I thought I would never want to see a uniform again, but the town of Washington, Connecticut, had shrunk a lot in the seven months I had been gone. As my nerves quieted down and the town's adulation for returning war veterans cooled off, my restlessness returned. I forgot the discomforts of Puerto Rico and yearned for the discipline and the camaraderie that the service had offered. I'd go to West Point, I told myself, but I quickly found that all appointments for the school year of 1899-1900 had already been made. However, I learned that a company of regulars was being formed for Philippine service. Since I already had satisfactory service in Puerto Rico, I was eligible to join.

This time I talked to my parents and got their consent to enlist. I made the trip to Camp Meade, near Middletown, Pennsylvania, and was sworn in for service with Company G, 19th United States Infantry.

I was now in the regular Army. My career as a military man had truly begun. This time, however, I enlisted under my right name.

Chapter Three

The Philippine Insurrection

The last military action of the Spanish-American War was the battle for Manila. Admiral Dewey had signed formal surrender terms in Manila on August 14, 1898, but this did not solve the problem of government in the Philippines. The collapse of Spanish power and the coming cession of the islands to the United States brought a rapid deterioration in Filipino-American relations. Emilio Aguinaldo, leader of the Filipino insurgents, had hoped and planned for recognition of a Philippine republic when the Spanish were defeated. Furious at the way his plans had gone awry, he fostered unrest among his people while the United States Army was trying to restore order.

On February 4, 1899, as the Treaty of Paris was about to be ratified by the United States Senate, Aguinaldo's followers rioted in Manila. The United States Army there consisted of about 21,000 men most of them volunteers about to return to the States. Under the command of Maj. Gen. Elwell S. Otis they controlled little more than the city of Manila. The insurgents held a semicircular line of blockhouses around the city with a force estimated at 40,000 determined Filipinos. In a two-day battle Aguinaldo lost about 3,000 men and the United States Army of Occupation 250.

It was this insurgent reaction that caused quite a stir in the States. A larger American force was needed to keep order. Within seven months 35,000 reinforcements were on the way to the islands, and I was one of them.

After about a month at Camp Meade, the 19th Infantry entrained for the West coast, and it was en route that I realized there was a sharp difference between what the public thought about the regular soldier and what I thought at that time. We had stopped at Lake Tahoe, Nevada for coal and water for the engine. Several buddies and I were walking along the station platform when two elderly women approached us with a tray of doughnuts.

"What unit are you boys with?" one of them asked.

"Nineteenth Infantry," one of us replied proudly. Our mouths watered for those doughnuts.

"Volunteers?" one of the ladies asked.

"No, ma'am, regulars."

The word "regulars" must have been a dirty word to those good people. They stared at us for a moment as if we had just run off with the town's bank funds, then turned and hurried away. This was my first experience as a second-class citizen. In later years, I learned that most people thought that regular soldiers were illiterates, thieves, and half-wits who had joined the Army to stay out of the clutches of the sheriff. In many towns women would sweep their daughters off the streets at the first sight of a soldier. Unfortunately, this attitude still prevails in peacetime in America, and it isn't helped by judges who tell a juvenile delinquent to "join the Army or I'll put you in jail."

We sailed from San Francisco aboard the *Tartar* in June, 1899, and arrived in Honolulu a week later. As we entered the harbor, I remember hearing the most enchanting Hawaiian song I have ever heard in my life drifting out across the water. A number of us rushed to the rails, expecting to see beautiful Hawaiian maidens heralding our arrival with their famous hulas. Instead, the music came from a boat going in the opposite direction. The people aboard were lepers on their way to a colony on the island of Molokai where they were to spend the rest of their lives. The song they sang was "Aloha," which to this day has a special meaning to me because of that pitiful, unforgettable sight.

We arrived in Manila about three weeks later and were encamped on the outskirts of the city. There we remained for about two months on police duty, helping to make Manila safe for democracy.

This was the "front" as far as we were concerned. Manila, like Saigon today, was still occupied by many insurgent guerrillas waiting for the right time to take the city away from the Yankees.

We were not advisers to anyone in those days, and we quickly learned that the major hot spots which had to be policed were the red-light districts, and the way we cleaned up those hot spots was simple and effective. All the girls and boys who snarled at us were promptly shipped over the front lines to the forces of Emilio Aguinaldo. Those who smiled and greeted us in a friendly manner were allowed to stay.

Upon completion of our police duty in Manila, my battalion was ordered to do guard duty around the Mariquina water works, about 15 miles outside of the city limits, and about five miles from the fighting line.

On the afternoon of our arrival we relieved a battalion of the 24th Infantry, a Negro regiment, which had been guarding the water works. I was ordered to take over the cossack posts, which were then occupied by a squad of the 24th.

As my squad of eight men made its way toward one of the posts, several Negro soldiers came running down the trail toward us. Their eyes were bulging with fright, and our first reaction was to crouch down, cock our rifles, and wait for the enemy, who must be right behind them.

Seeing nothing chasing them, I hollered to them to stop, but they kept on running at top speed without saying a word. We approached their abandoned post carefully, but still saw nothing. I shrugged. After all, the men were due for relief, and if they chose to leave at full speed and without saying a word of greeting as soon as we showed up, that was their business.

Ten minutes later we found out why the squad of the 24th had left so hurriedly. One of my men started to sit down on what looked like a log. It wasn't a log. It was ten-foot boa constrictor as big around as a telegraph pole. We promptly killed it with a couple of rifle shots. However, it was all I could do to keep my men on the job until we scouted around the outpost area and failed to locate any more members of the snake family.

Our task on outpost duty was to scout toward the insurrecto lines and assert ourselves to the villagers. We saw more alligators, snakes, and crocodiles than we did *insurrectos* or villagers. By the time we arrived at many small settlements the villagers were gone. We would poke around in the huts and leave.

These forays into enemy territory gave me an insight into the life of the Filipinos. The people lived primitively in grass huts, yet inside each would be some object of civilization that they obviously valued. For instance, many houses had Seth Thomas clocks, none of which worked, made in Thomaston, Connecticut, my home state, sitting on tables or hanging on the walls. Some had Spanish weapons and coats of mail, which they or their ancestors had stolen from their Spanish conquerors years before.

It was during one of those forays into enemy territory that I first encountered the water buffalo of the Far East, better known in the Philippines as the carabao. We were scouting through a narrow ravine with steep walls on each side, and a wet, slippery trail underfoot, when down the ravine came a wild-looking carabao, heading straight for us on the run. All of us scrambled safely up the walls of the ravine except Pvt. Allen S. Fletcher, who slipped and fell back to the bottom of the ravine. He landed right side up fortunately, coolly raised his rifle, and dropped the carabao in its tracks a few feet from where he sat. Fletcher was promoted to corporal for that neat little job, and later was made a commissioned officer in the Philippine Scouts, becoming famous for his work among the Moros on Mindanao.

We were relieved after an uneventful month, and our battalion was ordered to the island of Panay. The transport was the *Camp Dix,* which had just unloaded a cargo of horses from the states. As fast as the horses were taken off, we were loaded aboard and given their stalls—singles for enlisted, doubles for officers. Since we were always "sized," that is, lined up by size before marching off somewhere, I was invariably the last to arrive anywhere, and for once this was to my benefit. By the time I got aboard there were no horse stalls left. I found a spot on the open deck at the bow, and spent two days among the links of the anchor chain.

We landed at Iloilo, the capital of Panay, about the middle of November, 1899, and were told that our job was to help pacify the island. We were to sweep out in a fan-shaped deployment and cut across the island. There were supposed to be hundreds of guerrillas in the hills. Our job was to find and deliver them.

Our first day out we encountered opposition, and I had my first taste of enemy fire. Up until that time I had not taken the insurrection seriously, but when I saw the results of a clever ambush I realized that there really was a danger and that these people could, and did, kill Americans. We were actually surrounded by guerrillas, but seldom saw them. As we swept across the island, we learned that the people of the villages were docile but by no means sympathetic. Their loyalty was to Aguinaldo and his men, who watched, stalked, and waited for us. I realized that if all of the islands were like Panay, their pacification would be a long, costly, and patience-demanding project. If it was the American dream that these ignorant primitive people were ever to govern themselves, I felt then that it would take a generation or more.

While contacts with the enemy were few, and the toll of Americans light, I saw again what I had seen in Puerto Rico. A greater enemy stalked us—the enemy of disease. Lepers were everywhere. Smallpox, typhoid, yellow fever, malaria, and other tropical diseases were so common among the people that I wondered how any ever survived to become adults. The death toll was high, but it was even higher among Americans. There was a natural immunity of sorts among the natives, while the Americans had little or none to tropical diseases. Remembering the lessons of Puerto Rico and the lectures of my mother, I took my quinine regularly, boiled all the water I drank, and never ate native food. I was never sick a day in my five years of service in the Philippines.

There was little rest for us in the following months. We were kept on the move through the hills of Panay to prove the American presence and keep the guerrillas off balance. In January, 1900, our regiment moved to Cebu Island, where the hard-core remnants of the insurrection movement still held out. Our job was to search out, and kill, if necessary, those who opposed us.

Back in the United States, William Jennings Bryan was running for President. He and his followers believed the irresponsible stories that were spread in the newspapers about the atrocities committed by American troops against the defenseless Filipinos. I won't dignify the charges with explanations, except to state categorically that no Americans that I ever saw committed any atrocities, and the shots that rang out of the jungles and killed Americans did not support the statement that the Filipinos were defenseless. Besides, Americans don't know what atrocities really are, until they see what the Asians can dream up to do to a man.

Just as we were to take to the hills of Cebu, Sergeant Gunther called us together. “Men, they say back in the states that we have been committing atrocities against the natives here,” he said. “The orders are that no man is to fire unless fired upon. Any questions?”

One private from Texas looked at the sergeant quizzically. “Do you mean that if a Filipino comes out of the grass and levels his rifle at me I can't fire at him until he pulls the trigger?” he asked.

The sergeant looked at his questioner a long moment. “Your question proves you ain't deaf, soldier,” he said. “You heard me right. The other man fires first. Your job is to stay alive and get the second round off. It better be a good one.”

The Texan was quiet a moment, then drawled, “If it's all the same to you, Sarge, I'll play it like my daddy taught me back home. Any hombre that draws a bead on me is going to have to take his chances that I remember what you said. My daddy taught me to shoot first, and he taught me that before I ever met you. I sure do hope I can remember that you said I should let him squeeze his trigger before I let go. And I sure hope you don't get mad at me if I forget."

It wasn't long before we were fired upon so we could shoot back—officially. The first day that our battalion pushed inland toward Mt. Sudlon, the lead squad ran into a large band of insurrectos who fired from entrenched positions. The fighting raged for several hours as the regiment spread out and returned the fire. My squad was ordered forward. We held our fire as we inched up the slope through the thick elephant grass.

I couldn't see very far in front of me, but as I neared their trench, I could hear the Filipinos yelling excitedly as they started to run. I stood up quickly to take a look, and my hat flew off. Directly in front of me six natives were firing right at me and my squad. My rifle was firing before I could think about it. The native in my sights dropped, and my men accounted for the others. That night I filed the first of six notches on my gun.

When the skirmish was over, I retrieved my hat, and found it had a bullet hole through the crown a half inch from the top of my head. It was one time in my life I was glad I wasn't a full five feet six inches tall. If I were, I wouldn't be writing these memoirs. I've never objected to my height since that day.

Besides the hole in my hat, I collected the bolo that my late opponent had carried, and started a collection which I have to this day. I also picked up his rifle and felt a strange reaction as I examined it closely. It was a Remington, made in my home state of Connecticut, and I wondered by what devious route it had taken to get from there to that instant when it had been leveled at my head. I never found out, and never will.

The next day I earned my second bolo and my second notch in an identical situation. I was getting quicker on the trigger and a lot more savvy about positioning myself.

The charges of American atrocities grew in intensity in the States as we closed with the enemy half a world away. The mail we received was weeks old, but relatives and friends in the States were telling us what Bryan and his friends were saying about American atrocities, and they hoped the tales weren't true. They weren't.

My experience with an enemy atrocity was a vivid one. The day after the Mt. Sudlon skirmish we moved on in pursuit of the natives. Along the way one of the men in our company stepped off the trail to relieve himself. A few minutes later the company commander stopped the column and ordered me to take my squad back and look for the man because he had not caught up with us.

My squad searched for the missing man all afternoon. We saw no sign of him, and returned to the company bivouac area. The next morning we searched again. We found him this time. His body was in four separate, grotesque bloody pieces. His arms and legs had been tied to four carabao heading in four directions. The animals were whipped. The rest can be imagined. From the time I reported that act to the captain, the order to let the enemy shoot first was never repeated.

If there are those who believe that Americans commit atrocities, I can relate one instance which was typical of the way G Company of the 19th Infantry treated the enemy. It was a few days after the above-mentioned incident that we came across a Filipino boy and his mother along the trail. The boy was about fifteen or sixteen years old and had been shot in the leg. The wound was badly infected. Obviously, the boy had gotten it in one of our skirmishes.

Helpless and terrified, the lad fully expected to be shot. Instead the captain detailed several men to bandage him up and take him back to Cebu. It took several weeks for our medics to get him healed, but he was finally released and led back to his village. Our company took up a collection of $100 and gave it to his mother who had never thought she'd see him again.

Believe me, you did not need to speak the language to know what she was saying through her sobs of joy.

After the Cebu campaign I was promoted to sergeant. Our unit was assigned to garrison duty in the town of Naga in the fall of 1900 to begin what is today called a "civic-action" program. By this time most of the insurrectos had taken off their uniforms and hidden their rifles and bolos. During the day they would smile and wave to us as we filed through their villages, but at night they would uncover their weapons and seek out the American posts where they would kill and maim in the darkness. Every village was a haven for them, and every village had tunnels and dugout positions where they would literally go underground in the daytime.

The philosophy of winning this type of guerrilla warfare was to "civilize" by propaganda. The American forces offered a pardon and thirty pesos to anyone who would surrender and turn in his weapons. Only a few did so, and those few were organized into units known as the Filipino Scouts.

The guerrilla phase of the insurrection was tying down about 70,000 American troops by this time. In the States, the re-election of McKinley in 1900 caused a deeper look into the Philippine problem. The result was that a number of State Department and War Department experts arrived to study the people and figure out what it would take to make them stop fighting Americans and start on the road to self-government. It was our job to carry out the program dreamed up by these experts, which included the setting up of schools for the children wherever American troops could do so.

"Sergeant Foulois," my top kick announced one day with a big grin, "you are now a schoolteacher. Congratulations!"

I thought the sergeant had been in the Philippines too long. To tell Foulois that he was a schoolteacher was like telling one of our Army mules that he was the company commander.

"Sergeant," I told him, "the joke's on you. I never finished school. I couldn't teach anybody how to write his name."

The sergeant's grin was suddenly lost in the storm that

crossed his features. "That's exactly what you are going to teach, Foulois. Your job will be to teach every kid in Naga to write his name, and start from there. You're stupid, but you're a genius compared to those kids who have never heard of a school. Get going!"

I went. For the next eight months, using my "carabao" Spanish and the help of an interpreter, I not only taught those kids how to write their names, but English, history, geography, and current events as well. The events weren't very current because I used three-month-old newspapers, but I opened their minds a little, and let them know that there was more to the world than the village of Naga and the island of Cebu.

Although I scrupulously avoided religious matters in my teaching, the major opposition to my initial efforts came from the Catholic padres who saw in me a threat to their own teaching. They went to the parents and told them to keep their kids at home. But the kids came anyway. The need for what little education I could give them was great. The minds of those kids were like vacuums, waiting to ingest something besides hate, fear, and superstition.

The satisfaction I gained from my stint as a schoolteacher was more than I ever imagined it could be. There is nothing in my view that could equal the learning looks on the faces of those Filipino children. Years later I learned that many of them became leaders in their government, and led the guerrilla forces against the Japanese during World War II. I like to think that I gave them a start on the road to intellectual freedom.

In February 1901, I was given another order, and the conversation between my company commander, Capt. Truman O. Murphy, and me went something like this:

"I've got orders here sending you to Iloilo to take an examination for a commission," Captain Murphy said.

"But, Captain, I don't have the education to take any examination," I replied.

"Sergeant, you know better than to question orders. You either go or get a court-martial for disobedience of orders.

Which is it? Captain French of our regiment is president of the Promotion Board, and he knows your field record."

I went to Iloilo, but I felt miserable about having to take exams that I knew I would flunk. Captain French told me I could go to a school for two weeks to prepare for them, along with about twenty-five other enlisted candidates.

For the first time in my life I studied geometry and trigonometry, but deep down I knew that my newly acquired knowledge could never match that gained by my competitors in formal schools. The night before the exam, however, Captain French called me to his quarters.

"Foulois," he said, "I know you don't have any confidence in your ability to pass that exam tomorrow, but I want to give you some advice. Answer every question whether you know anything about the subject or not. Credit will be given for effort, but you can only be given zero for blank answers. Understand?"

I said I did and turned to go.

"One thing more, Foulois," the captain said. "Due credit will also be given for your field-service record. Do you understand what I am trying to tell you?" I nodded, thanked him, and left.

The examination next day was as tough as I had expected it to be. It had been made up by a West Pointer, and only a West Pointer could have passed it. But I took Captain French's advice, and answered every question. I returned to Cebu and waited.

While I had been gone, a vacancy had developed for a first sergeant, and when I returned, Captain Murphy asked if I wanted the job. There were three other sergeants who ranked me, but they had turned it down. I accepted.

It took about five months for the results of the commissioning exam to be announced. To my genuine surprise, I was on the list of new second lieutenants. I thus "crossed the line" and became an officer on July 9, 1901. There is no doubt in my mind that I didn't win my commission on the basis of the answers on the test. Whatever value they attached to my two years of field service with troops must have outweighed my ignorance, and I knew that this is what Captain French had been trying to tell me. I have no regrets, and have always thought that all officers, regardless of background or schooling, should have probational duty with troops during the first year of their service, whether before or after their commissioning.

The thirteen months between May 1900, and June 1901, were the most difficult of the insurrection. Fighting was bitter and brutal with no quarter given or asked on either side. More than 2,800 separate battles and engagements were fought, and the casualties on both sides were high. But the insurrection was essentially over when Aguinaldo was finally captured on March 23, 1901.

President McKinley appointed William Howard Taft chairman of a commission for the Philippines which was to legislate for the islands and prepare the way for the establishment of civil government. On July 4, 1901, civil government was officially begun. There was to be a gradual extension of autonomy with the Philippine Commission acting as the executive head of government and sharing the legislative powers with the Philippine Assembly which was elected on the basis of population. As the insurrection slowly flickered out, the Army troops took up civil government duties. Hundreds of civil administrators, doctors, teachers, and health experts arrived from the States.

In September 1901, two months after receiving my commission, I was ordered to Cottabato on the island of Mindanao where my battalion took up civil government functions. I immediately assumed ten jobs: inspector of customs, collector of internal revenue, captain of the port, city treasurer, chief of police, fire chief, topographical officer, engineering officer, signal officer, and commanding officer of Company D, 17th Infantry.

These were challenging tasks which left no time for

pleasures of any kind. And they led all of us into strange situations which gave us experience unique in American military history. One of these situations resulted in a burn-scarred hand for me, which is a constant reminder of those Philippine days.

Shortly after I assumed my ten duties, the battalion doctor reported an extremely high incidence of syphilis among our men. As police chief, I had to investigate where they were getting it. I knew that there is one man who should always know everything in an infantry outfit—the first sergeant. He did. He told me that our soldiers had reconnoitered the area around Cottabato and found that a tribe of aborigines lived up the Rio Grande River in the caves. For a few centavos our men were hiring natives to paddle them upstream and let them off near the village. They would follow the many paths through the ten-foot cogon grass, and for only ten centavos get themselves a case of syphilis that would put them out of action just as surely as an enemy bullet.

The problem of venereal disease has haunted military forces for centuries, and still does in spite of all the wonder drugs of today. Many philosophies have been brought to bear on the problem and no one has yet solved it. However, we think we came as near to it as anyone, although I will not comment on the right or wrong of it today.

After consultation with the commanding officer of our battalion, we decided to adopt a British solution to the VD problem. The British had concluded that they could never control the basic physical urges of their men, but they could control the circumstances under which their soldiers satisfied their desires. They set up their own carefully controlled houses where the girls were examined by a physician before being hired and then not allowed to accept any clients except British soldiers.

As the police chief, it became my duty to organize the logistics end of the operation. Our battalion commander arranged with the government of British North Borneo for the importation of a collection of Japanese prostitutes under the chaperonage of an experienced madam. The Japanese were noted for having the cleanest, most disciplined girls in the Far East, and they were easy to obtain.

While negotiations were going on in Borneo on our behalf, I had an old Spanish convent on the outskirts of Cottabato completely renovated and drew up a set of regulations for running a proper house. We decided that the price for enlisted men would be fifty cents and one dollar for officers. The men could use canteen checks as payment, but officers had to pay in cash. The girls could cash the checks at the canteen or take their worth in merchandise. The girls arrived in due time and were quickly settled in their quarters. The doctor checked them over, pronounced them fit for duty, and the operation began. The result was just as we had been told it would be. Not a single new case of syphilis developed.

We continued the official whore-house operation until we got orders to prepare to move out against the Moros in the spring of 1902. About a week before we left, however, a fire broke out in the girls' quarters. As fire chief, I rushed to the scene and started a bucket brigade. The fire was coming out of the roof, so I climbed up and had the men hand me the buckets which I splashed on the roof timbers. A sheet of flames suddenly blasted through the shingles and smothered my hand. I dropped off the roof as if I had been shot, and had to stand by, soaking my hand in a bucket, while my men eventually put out the fire.

Whenever I am asked how I got my scarred hand today, I always give a truthful answer. "In a government whore-house fire in the Philippines in 1902," I say. I don't know why it is, but I get no second questions along those lines.

The Filipino insurrection was mild compared to the difficulties we had with the Moros beginning in the spring of 1902.* They had been present in the Philippines before Magellan discovered the islands. They had originated on the Malay Peninsula and had migrated through the East Indies.

They were gradually converted to Islam by Mohammedan missionaries from India and Arabia. For more than five centuries they had raided and ransacked the coastal towns for plunder and slaves.

The Filipinos were terrified of the Moros, and rightly so. Filipinos used bows, arrows, spears, and blowguns as weapons, and fought mostly from ambush. The Moros used spears, but liked hand-to-hand combat where they could use the double-edged *kris*, the broad, heavy, single-edged *barong*, and the longer *campilan.*

The arrival of the Spanish did not alter the situation for the Moros much except that they had a new enemy to fight. And fight they did for years, so that the Spanish were confined in the southern islands to Zamboanga, a few strongly fortified posts, and a garrison on the mountain-girded Lake Lanao.

When the Spanish left the islands, the Americans negotiated a tentative treaty with the Sultan of Sulu, thinking that he was the sovereign ruler of the Moros. Gen. John C. Bates, the officer responsible for the Moro area, thought the best strategy was to let the Moros govern themselves to keep them from going to war against American troops while the insurrection was being put down in the northern islands. Unfortunately, the general was wrong. The Sultan controlled only about a third of the Moros, none of them on Mindanao.

The Moros were determined to live by their own code—criminal, barbaric, and violent by American standards. They believed in taking what they wanted by force whenever and from whomever they pleased. Slavery, piracy, murder, and thievery were a normal way of life.

As the Moros sallied forth and raided the coastal towns of Mindanao, a young captain was sent on the first field expedition against them in November, 1900. It was John J. Pershing, who had already made his mark among us, not only from his conduct as an officer but from his unusual perceptive powers which marked him

* Moro is the Spanish word for Moor.

for greatness even then. He had a different concept about winning the Moros, as evidenced by this description that he wrote of their character in a routine report:

> The most infinite combination of superstitions, prejudices, and suspicions blended with his character make him a difficult person to handle until fully understood. In order to control him other than by brute force, one must first win his implicit confidence, nor is this as difficult as it would seem... He is jealous of his religion, but he knows very little about its teachings... As long as he is undisturbed in the possession of his women and children and his slaves, there is little to fear from him. As a rule he treats his so-called slaves, who are really but serfs or vassals, as members of his family; but any interference with what he thinks is right regarding them had best be made gradually by the natural process of development, which must logically come by contact with and under the wise supervision of a civilized people.

My first real experience in dealing with the Moros came in the fall of 1901 when I was sent after a gang wanted for murder, robbery, and slave taking. Taking some friendly Moros along as guides, I led my company up the Rio Grande into the area dominated by Dato Piang, the local chieftain. I made extensive maps of the area, first along the rivers and then inland. In my capacity as signal officer I supervised the stringing of a telephone line between the blockhouse at Cudarangan and Fort Reina Regente. Once this was done, I decided to try the Pershing approach to Dato Piang, and made arrangements to meet him. He and his tribe had been quiet for a while, and I wanted to be sure they would stay that way. Suspicious at first, he agreed to meet me at the blockhouse which had been one of the tiny fortresses the Spanish had maintained. When he arrived with his retinue I talked with him through an interpreter for a while, and then decided to show him something he had never seen when the Spanish were there. I picked up the telephone we had just installed, and rang my men at Reina Regente. Then I got Piang to listen in while a Moro on the other end talked to him.

Piang heard the voice on the receiver, and immediately dropped it as if it had scorched him. He burst outside to see where the man was. When he couldn't find him, he was wide-eyed with puzzlement. Then I played my trump card. I had brought along an Ediphone which had just arrived from the States along with a selection of wax records. I put on one of the cylinders, cranked it up, and let the music roll out. That did it. In Piang's mind, the Americans were truly superior to the Spanish. They not only showed great courage by venturing out into the countryside, which the Spanish would not do, but they could make their voices travel great distances and play music without instruments. Data Piang was impressed and we became friends.

While the chieftains of the various areas could control their people as a group, they could not always control them as individuals. All Moros had a nasty habit of running amok. They became what we called juramentado, which is the name reserved for a Moro who has worked himself up to an insane frenzy and goes forth to kill all infidels or unbelievers in Mohammedanism until he himself is killed. He fully expects to be killed in the process, and he actually wants to be killed. But first he must kill unbelievers to assure himself a place in Seventh Heaven where he would have eternal youth and an unlimited number of concubines.

Just as they would kill for their religion, so would they kill for personal gain. On one of my expeditions into the Cottabato valley area to go after a small bandit gang wanted for murder, we wiped out the band in a brief skirmish except for the leader who was badly wounded. We brought him back to Cottabato for medical treatment. As soon as he recovered, we held a public trial to give the natives their first experience of American justice. I served on the court-martial that tried him, and the evidence was overwhelming: he was found guilty. Then, acting as police chief, I supervised the building of the gallows outside of our guardhouse. It then became my unpleasant duty to spring the trap on him. It was an ordeal to me. To him it was a pleasant experience. He was pleased because he was going to his Seventh Heaven, and since he was going there at the hand of an infidel—me—he was assured of eternal youth and all the beautiful women he could want for his harem. He died with a smile on his face.

Dato Piang was an unpredictable quantity in our area of operation, and I had several brushes with him that could have been fatal; but I sidestepped them all somehow without his becoming juramentado. But an opportunity came about on a trip to Malabang one day that was to make him my friend for life.

I had persuaded him to accompany a few of us along the coast to Malabang for exploration and map making. In the party of about twenty of his men was his son, Gumbay Piang. It had been raining heavily, and the river we had to cross was swollen. We Americans crossed with some difficulty, but the Moros would not follow. We waited on the other side and watched them discussing the matter. Suddenly, the boy bolted away and jumped into the water as if to show his elders that what the white men could do he could do.

The current was swift, and the boy couldn't swim very well. I could see he was going to miss the sand bar that my men and I had all steered for successfully. I jumped in and pushed one end of a stick toward him. He saw it, grabbed for it, and I held on to my end, but the current was so strong I was slowly being pulled off the sand bar. An enlisted man rushed out and pulled us both to shore.

Dato Piang was grateful and gave me a kris which he had specially engraved. Two wars later I learned that the lad I had saved became a dato himself and organized a guerrilla force to fight the Japanese. In February 1942, he sent a message to Secretary of War Stimson, pledging that his 20,000 "bolomen" would fight to the last against the yellow invader and "die for Americans and their country."

In the spring of 1902 my battalion was ordered to assist Pershing in his Lake Lanao expedition. He was pushing into areas where no white man had ever ventured and he was having trouble. Our job was to make the trails into the interior wider so that pack trains could get through first and then wagon trains. My Puerto Rican engineering experience came in handy as I supervised the building of bridges. One day as I was directing the felling of some trees for use as girders across a ravine, Pershing showed up riding his favorite mare. A tall, lean man, he was a typical cavalryman of that day, He never said much, and when he did it was neatly spiced with curse words appropriate to the time, place, and intelligence level of whoever was on the receiving end of those words. He stopped, watched a moment as my men fitted eight-inch logs into place, and then asked, "Lieutenant, what are you building that bridge for, a siege train?"

"Yes, sir, if that's what you want," I said, knowing full well that the engineering manual only called for four-inch logs for this job. "I just don't want to have to come back here a second time and build it again," I explained.

Pershing smiled his approval and rode on. Years later, when we were both generals and I served under him again, he reminded me of that bridge, and told me to apply that same philosophy in building an American air service in France "because we don't want to have to come back here a second time and build it again."

The Lake Lanao campaign ended for the 17th Infantry in June 1902, as we were replaced by fresh American units. The United States Army transport, *General Kilpatrick,* picked up our scattered units at all of our outlying posts in the southern islands and took us to the States by way of Manila and Nagasaki, Japan. We were certainly ready to go home. Many of the men were suffering from dysentery, malaria, and typhoid fever. Most of them recovered eventually; some needed years to get over the effects of these diseases. A few later died of dysentery because they could never shake the strain of the bug they had ingested.

After arrival in San Francisco, our regiment was split up and assigned to Fort Wright and Vancouver Barracks in the state of Washington. Since we had served three years on rugged overseas duty, we would normally have been able to look forward to at least an equal tour of stateside duty in comfortable quarters.

Life at Vancouver turned out to be a change of pace we had not expected. Officers and men had to go to school five days a week and it was the first theoretical training I received in my military career. Here I read Army regulations for the first time and learned about military law. It was in the law classes that I learned what a fraudulent enlistment was and that I had committed a crime against Uncle Sam when I enlisted in my brother's name in 1898.

My conscience bothered me about this newly acquired bit of knowledge and I decided to seek help from one of my senators, Orville H. Platt. I laid bare my sin in a letter to Senator Platt, and asked him to intercede on my behalf with the War Department. Weeks later I received official confirmation that not only was I off the hook for a fraudulent enlistment charge, but all of my Spanish-American War service would be credited to my own account without penalty.

The assignment to Vancouver was not all school work. We had several athletic teams and all officers were supposed to be light and fast. I was both. I played fairly well all season on the football team until the last game against an Indian school at Portland, Oregon. The Indians were all my size except that they all seemed to measure about a yard and a half across the shoulders.

In the first few minutes of the game the Indians probed our line for a weak spot and quickly found it—me. They rolled their plays around my end and over me so many times that I felt as if a herd of Army mules had galloped across my body. I don't remember most of the game. I never played football again.

School days ended for us in March 1903. Our instructor, Capt. Carl Reichmann, had all the attributes of the old Prussian Army officer, including mustache, ramrod back, the traditional dueling scar on his face, a stern demeanor, and an intense desire to make real professionals out of his students. He was rough but fair, and we learned a lot from him—sometimes directly—and sometimes not.

The night before graduation, Reichmann called us out of the barracks and marched us to the officers' club. Among our group were four West Pointers who had graduated the previous June. When we arrived at the club, Reichmann announced in his fractured English, "Ve are going to stay here for the night. In the morning, ve have terrain exercise. You vill not drink water, whisky, or soft drinks tonight!"

None of the older officers thought of disobeying the captain's order, but the four young officers could see no reason for it and quenched their thirst several times. For the next eight hours or so we stayed in the club, listening to his stories which were spiced with moral. There was much to be learned from this man and his stories, and I never forgot either. "Vork when you vork," he would say, and "play ven you play, but never mix the two. And don't mix your drinks!" This is good advice even in the Space Age.

At daybreak the next morning, Reichmann ordered us to the stables where we saddled up and awaited his instructions. "Today you graduate," he said. "But you must prove you are ready. Ve are going on combat problem for next four hours. You vill try to show me that you remember everything I show you."

He outlined the combat problem which required much map sketching, and we spent the next four hours galloping up and down the hills around Vancouver Barracks. The older officers were trying hard to show the four young West Pointers how good we were but those kids were sharp. They could ride, and they could make excellent sketches. They were determined to prove to us that four years on the Hudson made them superior to those of us who had come up through the ranks.

Reichmann followed us everywhere, making notes in his notebook. When he declared the exercise over, he gathered us together and gave us our grades along with a lecture on why each had earned what he did. The four West Pointers scored lowest of all of us. When they protested, he proceeded to give them another lesson.

"Many things to you last night I told," he said in his characteristic accent, his moustache bristling. "Many things of last night you did not remember today. The four of you took a privilege you vould not have in combat. You felt the urge of nature because of the liquids you drank last night and you each relieved yourselves during the problem. You vill relieve yourselves before the actions or you vill be dead men in combat!"

This was Reichmann's way of making a point and he made it with me. I suddenly remembered the man who had been pulled to pieces by the four carabao. All that man had done was step off the trail to relieve himself during a combat patrol. Those four West Pointers learned a lesson and it only cost them a low grade.

An amusing incident which occurred while on duty at Vancouver Barracks, Washington, gave me an opportunity to meet the man who was idolized by every member of our fighting forces, President Theodore Roosevelt.

In the spring of 1903, he was touring the country, making political speeches, which involved a visit to Portland, Oregon, just across the Columbia River from Vancouver Barracks. On the day of his arrival at Portland my company of the 17th Infantry was detailed for duty as his honor guard around the platform where he was to make his outdoor talk to the citizens of Oregon and Washington. After posting my detachment around the elevated speaker's platform, I posted myself and my first sergeant, John Madden, at the top of the platform's stairway, where we had a clear view of the passageway to the speaker's stand and a clear view of the crowd.

The President arrived, rushed up the stairway, and immediately launched into one of his famous anti-pacifist speeches, chiefly aimed at the chief pacifist of the time, William Jennings Bryan. The President had been speaking for about five minutes when I noticed a slight movement among the spectators near the front of the platform. A middle-aged man was making his way slowly toward the stairway of the platform. As he started up the steps, I saw he was carrying a small package in one hand. I immediately called out to Madden to stop the man. Madden stopped him, snatched the package, and threw it as far as he could over the heads of the spectators who were scattering in every direction.

The owner of the package was trying to explain that the package contained only an old family Bible he wanted to give to the President. Madden and I marched him over to where the package had landed, ordered him to pick it up and open it. Sure enough, all it contained was the man's old family Bible!

While our security action was taking place, the President, observing it all, never once let up on William Jennings Bryan until we brought the middle-aged man and his Bible up to the speaker's platform. Then he stopped, accepted the Bible, and thanked the donor and Madden and myself for our part in the action.

This was my first meeting with President Theodore Roosevelt. I met him again at Fort Myer, Virginia, in 1908, during our initial tests of the Baldwin airstrip and the Wright *Flyer.* On this second meeting I had an opportunity to recall our first meeting at Portland in 1903, which he recalled with a chuckle, stating that he still had the Bible.

No one knows exactly what happened, but a few weeks after graduation from school the 17th Infantry regiment received orders to return to the Philippines. There were two rumored reasons why we were going back when other stateside regiments had not yet been overseas since the insurrection had begun. One rumor had it that the 7th Infantry at Fort Thomas, Kentucky, was to go, but someone in Washington had made a mistake in numbers. The 7th was ordered to Alaska instead of the Philippines. The other rumor was that, being on the West coast, we were closer to the port of embarkation and it would be cheaper to send us than to send a unit from the East coast. Whatever the reason, we went and were assigned to Mindanao and Jolo Islands to continue the fight against the Moro juramentados.

Things hadn't changed in the islands during the fourteen months we had been gone, but the Army had learned a lot. We were now armed with better weapons. The Krag-Jorgensen rifle gave way in 1903 to the improved bolt-action magazine-type Springfield rifle. And the .45-caliber automatic pistol had been developed to replace the .38-caliber revolvers because we needed a sidearm that could stop the bolo-wielding Moros in their tracks, and the .38 couldn't.

The role of the military was the same, but the chain of command was different. The ranking military man now was Maj. Gen. Leonard Wood, a medical doctor who had served in Cuba as Teddy Roosevelt's superior when they were number-one and number-two men in the famous "Rough Riders." Wood commanded all troops in the southern islands but took his instructions from the civil governor of the Philippine Islands, William H. Taft.

The situation had altered considerably with the Moros. Formerly friendly Moro tribes had now turned against the Americans. Dato Piang, the chief I had befriended, had lost control of his people to his son-in-law, Dato Ali. This turn of events had been caused by the fact that when Taft arrived to take over the Philippine Commission he had declared a general amnesty and turned loose one of the biggest bunches of cutthroats the world has ever seen to return to their tribes and take up where they had left off. In addition, he imposed taxes on the populace they had never known before. The Moros, for example, were taxed for their bolos, knives, and other weapons. This tax was a supreme insult to them because their weapons were their badges of authority.

Any military man could have told Governor Taft that these two decisions would mean trouble, but the military was not consulted. It still amazes me that the representatives of our government could not remember that "taxation without representation was tyranny" and had been one of the reasons for our own American Revolution.

My first jobs after arrival were as post commissary and quarter-master officer at Zamboanga. My first real combat mission didn't come until the summer of 1904 when I acted as topographical officer for General Wood and went on an expedition to capture the Sultan of Taraca, known as Macui. The Taraca Moros had an exceptional record for killing American soldiers, stealing arms, and otherwise harassing our forces. General Wood in March, 1904, wrote in his official report:

> Conditions on the lake [Lake Lanao] have been and are intolerable… It is unsafe for a man to venture a few yards from Camp Vicars… The Taraca Valley is the only section of the Lanao country that is openly hostile and defiant… Peaceful means have been unsuccessful… Captain Pershing, on his last expedition, punched a hole through the Taraca settlements, destroyed a couple of cottas, but the greater portion are untouched, defiant, and intensely hostile… I intend to visit the Taraca Valley peacefully, if possible, with a force of troops, and establish a camp near the mouth of the river.

The expedition of 1,000 men left in 1904, and we battled our way inland through a well-planned series of cottas and forts for the next two weeks. We were ambushed several times, and lost about a dozen men. We killed twice as many Moros, and destroyed their fortifications whenever we found them.

I earned my third, fourth, and fifth notches during this campaign. The third was occasioned by an assault on a small blockhouse surrounded by a moat. A log laid across the moat was the only way into the blockhouse.

There was no sign of life as my patrol approached, but I was suspicious. I stood up, but didn't draw any fire, so I inched my way across the log, followed by my men. I cautiously pushed open a door with my .38 and was met with a swishing sound which was occasioned by a Mora swinging his *campilan* at my arm. I pulled back my arm, and the huge blade slashed into thin air. The only piece of me it caught was the tip of a finger.

Three revolvers, including mine, opened up on the Moro at pointblank range, and he fell through the doorway on top of me. While I was disentangling myself from the dead Mora, the rest of my men forced their way inside the blockhouse where a dozen more Moros waited. A few surrendered. Those who didn't were shot in the melee.

Several days later my patrol was ambushed by a half-dozen Moros. One of them leaped out of the tall grass at me with a kris in one hand and a knife in the other. I emptied my .38 into the hurtling form as my men reacted against the others, wounding two and killing two. One fled screaming through the brush.

When the encounter was over, I found that the Moro I had shot was a woman. Since Moros of both sexes dressed and wore their hair alike, it was hard to tell man from woman. All I can say is that they fought like men so they had to die like men. This woman's kris also decorates my wall today.

Notch number 5 came a week or so later in another ambush as we moved into the interior against the Sultan of Taraca's hideout. We were advancing through tall grass toward a suspected enemy position. I could hear movement in the grass in front of me, but couldn't see anything. Since I was at the point of my patrol, the noise couldn't be anything but Moros. I stopped, and the noise got closer. I cocked my revolver and pointed it toward the noise. The grass parted, and I was face to face with the toughest, biggest Mora I had ever seen!

There was a split second of silence, and then we both reacted. He raised his *campilan* over his head with both hands and I fired. The bullet caught him squarely in the center of his forehead, and he collapsed. As he did, I felt a sharp sting on the side of my face, and thought I had been nicked by a bullet. I hadn't. I had been hit by a poison dart from an unknown adversary. To this day I don't know where it came from.

When the skirmish was over, my patrol dropped back after collecting the weapons of the dead Moros and dressing our wounded. I was worried about the poison-dart wound, but got one of my men to wash it out with raw whisky. I never did have any reactions from the poison, but even today, more than sixty years later, the skin over that area of my face comes off when I shave. So every morning of my life I am reminded of the day I earned my fifth notch.

But the Army had reason to remember that day, too. The *campilan* I had "liberated" was a beautiful weapon of finely forged steel, and sharper than the finest razor. The handle was beautifully carved, and the blade etched by a craftsman. When I got back to camp with it, one of General Wood's staff officers looked it over.

"Foulois," he said, "that weapon is the finest I have ever seen. I think you should give it to the general for a souvenir."

"I think the general would rather go out and get his own souvenirs," I told him. After thinking it over a little while, however, I thought I might give it to the general, when one of my men told me why the weapon was so valuable. It had belonged to the Sultan, the object of our whole campaign!

Sure enough, I had killed the Sultan of Taraca and had captured his badge of office. For a long time I slept with his *campilan* in my bed with me to keep it from being expropriated by overzealous aides to the general. It is the finest piece of cutlery in my collection today.

The death of the Sultan solved the Taraca uprising, and our troops returned to Zamboanga. The next expedition I took part in was in May, 1904. Sixteen men of F Company had been ambushed and massacred at Simpetan, 40 miles from Reina Regente. The patrol led by Lieut. Harry A. Woodruff had been sent up the Rio Grande River to see if they could locate Dato Ali, now the chief troublemaker in the area. Ali's force of 160 men had laid a trap with great care, and Woodruff's company was overwhelmed.

Throughout the remainder of 1904 the hunt for Dato Ali continued, and I did my share of patrolling. Because of losses owing to the Moros and the expiration of enlistments, our company was down to about 50 men. As a result, "provisional" companies were formed for war duty and I was attached to the 17th Provisional Company. I went on expeditions to Lake Liguasan and Gata without incident and was able to dodge wild bees, crocodiles, snakes, malaria, dysentery, diarrhea, and yellow fever successfully.

In May, 1905, I was ordered to take a provisional company of Moro Scouts and Philippine constabulary to Jolo Island to find a band of Moro outlaws led by Data Pala, who had recently run amok in the town of Lahud Datu and killed a number of its inhabitants.

Previous attempts to capture Pala had been unsuccessful, and the Governor of Jolo, Major Hugh L. Scott, had asked General Wood for help. My contingent arrived on May 1, and we were in action within a few hours. For the next ninety-six hours we battled the hostile Moros and drove them back. On May 5, elements of the 14th Cavalry, the 17th Infantry, and my Scouts surrounded a fortified renegade position on Mt. Talepao, a huge volcanic mass rising steeply out of the jungle. The 14th and 17th sent their men up the slope from the north, and we charged from the south, dodging rifle fire all the way. As my Scouts closed in, they were firing at the Moros at close range. One of my men fired point-blank at a Moro and killed him, but the Mora had already raised his barong and my man didn't duck. The barong came down on the Scout's head and split it open clear down to his shoulders.

The skirmish was over quickly. The score was four of our men killed and six wounded. Twenty-three renegade Moros were killed. One of them was Data Pala.

Three days after this engagement, an ally of Pala's named Data Hati and a large number of hostile Moros had holed up in a big volcano crater near the eastern end of Jolo Island. General Wood, taking personal charge of the operation, ordered our units to reconnoiter their position.

The crater was adjacent to the shore line, and its seaward wall had long before dropped into the sea. The route in was covered with dense tropical jungle growth. If we advanced through the jungle, we would be under enemy fire from the top of the crater. The one path to the top of the crater had been booby trapped with poisoned, sharpened bamboo stakes and entanglements.

General Wood called the officers together, and while we

were discussing the situation, word came that Data Hati wanted to surrender and had ordered all the booby traps dismantled. The general ordered two companies along the rim trail to accept the surrender. Slowly and cautiously, about 450 Moros—men and women—came in a single file down the trail as we covered them with every gun we had.

They were dressed in their finest and brightest turbans, sashes, and heavily embroidered jackets as they always were when they made up their minds that they were going to die fighting.

The men quietly piled their rifles, spears, barongs, and krises in a heap when they reached the bottom. They were completely docile and seemed to expect us to shoot them down.

When the last man laid down his weapons, General Wood strode forward and talked to them through an interpreter. He told them not to fight the Americans any more and to go back to their homes in peace. When his words were translated, a great joyful shout rang out. They *had* expected to die, but it was the first time any Moro band had fully prepared for a fight and then surrendered without one.

It was many weeks before we learned the answer to this riddle. We found that the night before we had arrived at the foot of Hati's crater two United States destroyers and a gunboat had anchored offshore. All night long they had focused their powerful searchlights into the open side of the volcano, bathing the whole area in a dazzling glare. To the Moros, those lights looked like giant suns coming out of the darkness. If the white men who pursued them had the power to turn night into day, they knew they could not hope to win against such magic.

I don't think the Navy ever realized how they had won the fight against Dato Hati for us or their publicists would certainly have played it up. Maybe they can, now that I have told the story for them.

Philippine service for the 17th Infantry came to an end in July 1905, although the insurrection was not over. The Taft Commission, doing their best to bring law and order to the islands, devoted their attention to the problems of reconstruction. But in many provinces the local Moro leaders refused to follow their leaders and submit to white rule. They continued their raids and pillaging until late in 1906. The Army's job was to keep the pressure on these renegades until they were either decimated or realized that the Americans were there to stay and would not leave until the job was done.

Governor Taft left the Philippines in January 1904, to become Secretary of War under President Theodore Roosevelt, but he was followed by other governors who gradually relinquished their hold on the government. Representative government was introduced, the Philippine Commonwealth was established, and the process of orderly management of Philippine affairs passed slowly to the people under the leadership of Manuel Quezon, Sergio Osmeña, and Manuel A. Roxas. However, it was not until July 4, 1946, that the Philippine Commonwealth became an independent republic. Thus, it took almost half a century of American cooperation to give the Filipinos the freedom they wanted and deserved. This proves that the Asians do not learn about our way of government easily. For those who believe that our involvement in Vietnam and other countries in Asia will end soon, they need only to study the history of the Philippines. The Moros we fought in the 1900s were not organized as are the Communist forces of the world today, for the Communists are sustained by a firm, unyielding ideology and led by determined, competent leaders. Our struggle in the Far East is much more serious today, and far less likely to be resolved during this century.

There were many lessons learned during our Philippine service. For example, we found that a few hundred natives living off their land and fighting for it could tie down thousands of American troops, have a serious impact on the economy of the United States, and provoke a segment of our population to take the view that what happens in the Far East is none of our business. The Army also learned valuable lessons of leadership and logistics in the islands, but was not able to convey the theme of military preparedness to the American people. This was a pattern that I was to see many times over in my next thirty years of service.

Chapter Four

'Lighter-than-air' Days

I had made application to attend the Infantry and Cavalry School at Fort Leavenworth, Kansas, before leaving the Philippines, and the application was approved. I spent a thirty-day leave at home in Connecticut in the summer of 1905, and reported for school in September.

Adjusting to the life of a full-time student was not easy for me. With three wars behind me, I found it difficult to accept the theories about how to fight wars from books and from instructors who had never heard a shot fired in anger. But I tried to overcome my feelings and studied hard to come up with the right "school solutions."

I spent time at the officers' club and got acquainted with a number of officers who were attending the Signal Corps School there. Their description of their work in communications fascinated me, and I listened to them discuss their school problems for hours on end. They were not only interested in telegraphy and codes and other forms of signaling, but were exploring the use of free and captive balloons and airships which could be steered and were powered by gasoline engines. But most fascinating of all were their discussions about the work of a couple of brothers by the name of Wright, who had successfully mated the gasoline engine and the glider and made what they called an "aeroplane." I had never seen any of these air vehicles, and it was to be many months before I did.

I wanted to transfer to the Signal Corps, and requested permission to attend the Signal Corps School the following September. My application was approved.

I was sent to Wyoming for the month of August, 1906, to act as an umpire during maneuvers and begin the Signal School in

September. One month later, however, I was ordered to rejoin my regiment then being sent to Cuba with the promise that I could return to school the next year.

Much as our government had hoped, the signing of the peace treaty with Spain in December 1898, did not mean that Cuba was automatically able to stand on her own feet. Cuba had been set free of Spanish rule on the single condition that "she establish a stable government, capable of maintaining order and observing international obligations."

During the American occupation of Cuba, a military government had charge of the island until May, 1902, when control was relinquished to the Cubans. In 1906 mutterings of discontent were heard following national elections, and on August 19, 1906, the United States had to send in Marines to protect American interests. The president of the republic resigned, and the United States had no choice but to send in more armed forces and install a provisional government. The Army of Cuban Pacification was thus formed, and I became a part of it.

My previous mapping experience caused me to be assigned to Maj. Mason M. Patrick, chief engineer for the expeditionary force. He assigned me to an area around Ciego de Avila, located 280 miles east of Havana and 260 miles west of Santiago. My instructions were to make accurate military maps of my area and "other duties as may be assigned."

In May 1907, I was promoted to first lieutenant and ordered back to Fort Leavenworth to await the opening of the Signal School in September. I tackled my signal course with vigor, and I felt as if I was way ahead of my classmates because I had spent so many hours discussing the course with the class of 1905-1906. I became thoroughly intrigued by what little material there was on dirigibles and the things called aeroplanes. As part of the course, each student was required to write a thesis on any Signal Corps activity in which he had a special interest, so I decided to write mine on "The Tactical and Strategical Value of Dirigible Balloons and Aerodynamical Flying Machines."

My classmates and instructors thought I was some kind of nut, especially when I predicted that although "the dynamical flying machine is still in the experimental stage, if it continues to progress toward perfection as rapidly in the next year or two as it has in the past two years, we will see aerial navigation reduced to a very simple problem, and tacticians will have to revise their books on tactics…"

I also noted that France and Germany had "taken the lead in aerial accomplishments and England is also coming to the front." But my contemporaries really choked when I wrote that "our government has finally awakened to the fact that aerial navigation is no longer a theory, and if not handicapped by a shortsighted Congress, there is no reason to doubt that in a few years we will be abreast, if not ahead of, France and Germany."

Since we had been encouraged to think differently and imaginatively, I did not let the conformist snickers of my classmates deter me. I forged ahead in the development of my thesis and got bolder as I went along. Borrowing from the Bible and Jules Verne, I made some forecasts which I was able not only to see come true but to play a part in making them come true. Here are some excerpts from my paper:

> In all future warfare, we can expect to see engagements in the air between hostile aerial fleets. The struggle for supremacy in the air will undoubtedly take place while the opposing armies are maneuvering for position, and possibly days before the opposing cavalry forces have even gained contact. The results of these preliminary engagements between the hostile aerial fleets will have an important effect on the strategical movements of the hostile ground forces before they have actually gained contact.
>
> The successful aerial fleet, or what remains of it, will have no difficulty in watching every movement and disposition of the opposing troops, and unless the opposing troops are vastly superior in numbers, equipment, and morale, the aerial victory should be an important factor in bringing campaigns to a short and decisive end.

I predicted the obsolescence of the horse cavalry for reconnaissance, and said that "a modern military aeroplane" would replace it. I also predicted that the flying machine could not only reconnoiter the territory in front of an army more thoroughly and in a fraction of the time "but could photograph all of its main features and have the finished prints in the hands of the commander in chief in a very short space of time."

Emboldened by this venture into the world of my imagination, I predicted that searchlights and flares might be used by dirigibles and planes to discover night attacks on the outpost lines of ground forces below. I particularly foresaw the value of the aerial fleets in maintaining communications between elements of armies, for up-to-the minute reconnaissance of enemy positions, and for artillery spotting to improve the effect of its fire power.

To cap my bold venture into the theories of future employment of air vehicles, I felt that some means of communications between planes and between the air and ground were essential and that "the only means of communication are the wireless telegraph and wireless telephone."

I had studied the reports of Signal Corps experimenters and others and reported as follows:

> In the few experiments in receiving wireless messages in a free balloon, no difficulty has as yet been encountered. As regards sending, no attempts have as yet been made by our aeronauts, due to the danger of sparks being thrown off from the antenna, which, while being used in sending, carries many thousand volts. There can be no doubt, however, that this difficulty, as well as all others incident to the practical application of wireless in both sending and receiving, will soon be overcome.
>
> A Frenchman by the name of Berjonneau has perfected an instrument by which photographs may be transmitted by wireless over a considerable distance. If this instrument can be relied upon, and the wireless telegraph made applicable to

> aerial craft in flight, the aerial fleet of an army will not only be invaluable in securing data of the country over which it passes, but will be able to transmit at once by wireless photographs of the area passed over.

My concluding remarks contained more evidence of how far my imagination had been allowed to stray:

> In the preparation of this paper, no attempt has been made to set forth the value or uses of an aerial fleet in conjunction with the Navy or in combined Army and Navy operations. The use of the flying machine will undoubtedly be an important factor in future naval operations and the naval authorities are fully alive to its tactical and strategical value.
>
> In conclusion, I wish to remark upon the great interest that is being taken by the War Departments of France and Germany in the perfection of the dirigible balloon and the dynamical flying machine. Such activity on the part of these powers makes it imperative that we take a like interest if we do not wish to be outclassed in army equipment. Our geographical position with respect to Europe has saved us from numerous difficulties in the past, but it is hardly possible that we can expect to be unmolested in the future, with such means of transport and communications as are available at the present day.
>
> Our War Department evidently realizes the necessity of an aerial fleet, and if the air machines now practically completed are able to accomplish everything that is required in the War Department specifications, the United States will have outstripped its European rivals.

My thesis was dated December 1, 1907. It was read by the school faculty and sent to Washington where it was logged in as Document No. 15110-166 and passed to the staff of the Chief Signal Officer for study. I make a point of this because many years later the world was to give credit to a brother officer of mine named Billy Mitchell for being the first military man to foresee the future of the airplane as a weapon of war. With all due respect to Mitchell, neither he nor his worshipers can document that claim.

As far as I know, I was the first military man in modern history to predict the future military uses of the airplane and predict them accurately. Admittedly, I was not so sure of myself as Mitchell was a decade later in his tirades for the benefit of the press. After all, I had not yet seen an aerial vehicle of any kind, but if the vehicles were as good as the reports I read claimed them to be, I felt sure that flying machines would revolutionize the conduct of war. Proof of my claim is my thesis which was available to Mitchell when he finally came around to my way of thinking after World War I started.

In 1913, for example, Billy had written that at best aviation was a reconnaissance device and "the offensive value of this thing has yet to be proved." He added that "it is being experimented with—bomb dropping and machines carrying guns—but there is nothing to it so far except in an experimental way." He also made clear his own personal lack of interest in aviation by insisting that it was a young man's pastime and "no man should be taken in who is more than thirty years of age or married to start with."*

My thesis antagonized my cavalry brethren as I knew it would. I was attacking a sacred cow by saying that the Army's horses might be used only for ceremonial purposes someday as soon as flying machines were perfected. And to be so "way out" in my thinking as to believe that photographs might be transmitted from the air to the ground also branded me as a "Section 8" in the eyes of my Infantry colleagues. But I had transferred to the Signal Corps now by virtue of graduation from its school, and I didn't care what my contemporaries in the other branches thought.

There had been some significant lighter-than-air developments in the Signal Corps between the time of my enlistment in 1899 and my graduation from the Signal School in 1908. A Balloon Detachment had been organized at Fort Myer, Virginia, in May 1902, but outside of participating in maneuvers in Connecticut that

* It is my intention to set the record straight more than once in regard to Billy Mitchell in these pages. A zealous and jealous man by nature, it was inevitable that the two of us would clash someday. We did, and he attacked me viciously many times in print. For the first time I am going to tell my side of the story in the interest of historical accuracy.

autumn nothing much happened because of the indifference of the Congress and a lack of enthusiasm on the part of the Army. But the interest in ballooning as a sport, and the establishment of the Aero Club of America in 1905 influenced the staff of the Chief Signal Officer in Washington to take a new look at balloons and dirigibles.

Lieut. Frank P. Lahm, along with Lieut. Henry B. Hersey, entered and won the First International Gordon Bennett Balloon Race in September 1906, with the *United States.* Lahm had learned about ballooning from his father, Frank S. Lahm, in Paris, while on duty there at the famous French Cavalry School. His father was the Paris representative for an American corporation and had previously qualified as a spherical balloon pilot in 1902. The success of these two American Army officers focused attention on ballooning in the States as nothing had before. The Aero Club of America immediately sponsored a demonstration at Pittsfield, Massachusetts, the next month, and invited the Signal Corps to send two representatives as observers. Maj. Samuel Reber and Capt. Charles DeF. Chandler were designated and made their first ascents in the *Centaur,* piloted by Charles T. Walsh. The reports of these two officers upon their return to Washington caused the Chief Signal Officer to order a new spherical balloon late in 1906, the tenth purchased by the Army since its first try at ballooning during the Civil War.

While the Signal Corps was showing increasing interest in balloons, it had also decided to investigate the dirigible and the airplane of the Wrights in the summer of 1907. Because of this increased interest, Maj. George O. Squier, executive officer to the Chief Signal Officer, recommended the establishment of an aeronautical division. This was accomplished on August 1, 1907, when Brig. Gen. James Allen signed an office memorandum and ordered that the division "will have charge of all matters pertaining to military ballooning, air machines and all kindred subjects." Captain Chandler was detailed as officer in charge and Corp. Edward Ward and. Pvt. First Class Joseph E. Barrett were assigned to assist him. Thus, the United States Air Force was born.

Shortly after the establishment of the Aeronautical Division, a large number of inquiries were received concerning the Army's advertisements for proposals for an airship and an airplane. Frank Lahm returned from Europe with much valuable information on French, British, and German aviation activities and was promptly assigned to assist Chandler in assessing the proposals so they could come up with a set of government specifications for both types of vehicles. On December 23, 1907, General Allen signed Signal Corps Specification No. 486 for a heavier-than-air flying machine, asking for sealed proposals to be received by noon of February 1, 1908. On January 21, 1908, he approved Specification No. 483 for a dirigible balloon with a proposal deadline date of February 15, 1908.

The airplane had to be designed "so that it may be quickly and easily assembled and taken apart and packed for transportation in army wagons." It had to be able to carry two people having a combined weight of about 350 pounds and enough gasoline for a 125-mile flight. A speed of 40 miles per hour was to be averaged, with a bonus of 10 percent for each mile above that speed up to 44 miles per hour and penalties of 10 per cent for each mile per hour less than 40.

In addition, a one-hour endurance flight was required, and the airplane shall return to the starting point and land without any damage that would prevent its immediately starting upon another flight. It also had to be "sufficiently simple in its construction and operation to permit an intelligent man to become proficient in its use within a reasonable length of time." Two men were to be instructed in the handling and operation of the machine at no extra charge to the government.

The dirigible balloon had requirements appropriate to the nature of a steerable gas bag. It also was required to carry two persons having a combined weight of 350 pounds, plus 100 pounds of ballast "which may be used to compensate for increased weight of balloon while operating in rain." Speed required was 20 miles per hour with 15 per cent bonuses and penalties for each mile per hour above or below that figure. The speed run was to be over a measured course "between two and five miles against and with the wind." The time was to be computed "by a flying start, passing the starting point at full speed at both ends of the course." A two-hour endurance flight was required, and "the engine must have suitable cooling arrangements, so that excessive heating will not occur." A manometer and "all other fittings and appurtenances… required for successful and continuous flights" had to be furnished. In addition, "the scheme for ascending and descending and maintaining equilibrium must be regulated by shifting weights, movable planes, using two ballonets or other approved method. Balancing by the aeronaut changing his position will not be accepted." Two men were also to be trained in its operation by the successful bidder at no extra charge.

General Allen had been particularly impressed with the dirigible after attending the St. Louis Air Meets in October 1907, where a number of private individuals had exhibited their craft. One of these was Thomas S. Baldwin, who made his living performing such spectacular stunts as parachute jumps from balloons and had constructed his first dirigible in 1904. Using a light two-cylinder engine built by Glenn H. Curtiss, he carried off all the prizes with his *California Arrow* at the St. Louis World's Fair that year. By the time of the 1907 meet, Baldwin had established such a good reputation at home and abroad and his performance so impressed General Allen, that he discussed with Baldwin the practicality of building a nonrigid airship for the Signal Corps.

Since no funds were available for this purpose, Allen had to apply to the War Department's Board of Ordnance and Fortifications for $25,000, "or so much thereof as may be necessary" to purchase an experimental nonrigid dirigible balloon. His request was approved on November 7, 1907.

The bids received for the dirigible were opened on February 15, and on the twenty-fourth General Allen announced that of the nine bidders who believed they could meet the specifications, Baldwin's bid of $6,750 was lowest. He had 150 days to deliver the airship and meet all specifications.

The arrival of my thesis on dirigibles and flying machines in the office of the Chief Signal Officer caused me to be assigned to membership on an aeronautical board to conduct the airship and airplane trials after I graduated. Accordingly, I was assigned to the office of the Chief Signal Officer of the Army in July, and placed in charge of the Balloon Detachment at Fort Myer, Virginia.

The component parts of Baldwin's airship arrived at Myer during the week of July 20, and I watched Curtiss and Baldwin put it together. It weighed 253 pounds and had a gas capacity of 20,000 cubic feet. The envelope was made of two layers of Japanese silk with a layer of rubber between them. The car beneath the bag was 66 feet long and was composed of four longitudinal spars of Oregon spruce separated by small struts and held together by brace wires.

The engine on this airship was water cooled with four cylinders and generated between 20 and 30 horsepower. Mounted about one-third the length of the car from the front, it drove a tubular steel shaft on the front end of which was a single wooden propeller that had been designed by Lieut. Thomas E. Selfridge.

At the rear of the car was a fixed stabilizing surface back of which was located the vertical rudder for directional steering. Sitting in a canvas seat about two-thirds of the distance from the front of the car, the rear pilot operated the rudder. The forward pilot, seated immediately behind the engine, operated the altitude control with a fore-and-aft rod which, by changing its angle, controlled hinged superimposed planes resting on the car. The car was attached just below the bottom of the envelope by short suspension ropes, and a rope net of square mesh wrapped over the entire envelope supported the car.

I was introduced to Curtiss and Baldwin along with Lahm

and Selfridge. Since I was more interested in the engine, I spent most of my time with Curtiss, a serious man who had nothing but engines on his mind. When I reported to him for duty, he took one look at my clean uniform and with an air of exasperation suggested that I equip myself with five items if I expected to work with him: a suit of coveralls, a pair of pliers, a screw driver, a handful of cotton waste, and a bar of soap.

By August 1 the airship was fully inflated and rigged and the engine tested. Various adjustments of the rigging control devices and engine were made during a number of captive ascents. I asked hundreds of questions of Baldwin and Curtiss, and probably made myself obnoxious to them looking over their shoulders every time they tightened and tinkered and turnbuckled

The maiden flight was made on August 4 and lasted only seven minutes. During this and the subsequent test flights Baldwin worked the rear controls while Curtiss managed his engine and worked the elevator controls from his front position. They continued their test flights for more than ten days, and finally announced they were ready for the speed trials on the fourteenth. The course from Fort Myer to Cherrydale measured four and one-eighth miles. After an abort because of high winds in the morning, the two men went aloft late that afternoon and covered the course in seventeen minutes to compute an average speed of 19.61 miles per hour. This 0.39 difference caused 15 per cent to be deducted from Baldwin's bid price of $6,750.

The next day Baldwin and Curtiss fulfilled the two-hour endurance requirement by making eight round trips over the measured course. Their average speed was computed at 13.75 miles per hour.

On the eighteenth, the board, consisting of Majs. George O. Squier and Edgar Russell, Capt. Charles S. Wallace, Lieut. Frank Lahm, and I notified Baldwin officially that his airship had passed all of the flight tests and the only obligation left was to train two pilots to fly it. Instead, Baldwin and Curtiss offered instructions to the three of us who had gotten our hands dirty helping them get their ship in the air: Lahm, Selfridge, and me.

Baldwin selected me as the first to go aloft. He flew from the rear while I flew in front as the engineer-pilot. It was my first flight aloft in any type of aircraft and I was exhilarated beyond words. The flight lasted only about ten minutes, and I was too busy reacting to Baldwin's commands to think much about being above the ground with only a bag of air holding me up and four Oregon spruce bars held together by wire holding me in. But being air-borne, with the controls in my hands and the hot engine blasting me in the face, sent a surge of joy through my whole body that defied description. As we chugged around the Fort Myer parade ground, I looked eastward to the Capitol, the Washington Monument, and the Lincoln Memorial glistening in the bright morning sun. It was one of those sights a person never forgets.

After Baldwin and I maneuvered *Dirigible No. 1* to the ground, I started to get out, but Baldwin yelled for me to get back in. "You're not through yet, Foulois," he hollered. "It takes more than one take-off and landing to make a pilot."

We made two more wide circuits of Myer and then Selfridge and Lahm took my place as engineer. Then we alternated on subsequent flights in the rear seat operating the directional controls until we each had about a half-dozen takeoffs and landings.

One afternoon when Selfridge and I were scheduled to fly together, we decided that we would take our first cross-country flight to the Washington Monument, around the White House and back. After takeoff, with Selfridge in the rear, we headed out across Arlington Cemetery at about 300 feet altitude. Just as we reached the Potomac River, the engine backfired and black smoke poured out of it onto my face. I thought it was going to catch on fire. I signaled to Selfridge to go back to Myer.

On the ground, Baldwin was getting agitated and thought that when we had disappeared from his view over the cemetery that we had crashed, and he had not yet collected his check from the government. He made no effort to conceal his elation at our safe return. Selfridge and I made no effort to conceal *our* elation at getting back safely. Neither of us liked the idea of being 300 feet off the ground with an engine blazing away only inches from the highly inflammable hydrogen.

While we were conducting our tests, a quiet man named Orville Wright arrived at Myer with a number of boxes and crates which contained his *Wright Flyer* and started to assemble it. It took him several days to put the frail contraption together along with its starting track and its wooden tower for hoisting the catapult weight. A small open-end shelter was constructed on the east side of the parade ground to shield the airplane from the hot summer sun.

Between dirigible flights I watched Orville and his mechanic as they worked. They paid no attention to anyone else, and shrugged off all questions from onlookers. They talked only to each other, as if they were on a desert island miles from civilization, As their glider took shape and they mounted the engine on the frame, I wondered how such a combination of cloth, wires, pulleys, chains, and wood could ever carry two people for an hour at the fantastic speed of 40 miles per hour. But there was something about the way these men handled tools and went about their business that fascinated me. After all, I told myself, Orville and Wilbur Wright had been the first to fly a motorized glider successfully five years before, and had amassed a total of about twenty hours of flying time between them—more than anyone else in the world. If they said they could meet the specifications for an airplane then my money was on them.

The Wrights had not had an easy time getting to that point of proving their invention to the government. They had received enquiries from a foreign government in 1904 relative to the purchase of their machine, but they wanted to give the United States first chance at it. They wrote to the congressman from the Dayton district, R. M. Nevin, telling him of their accomplishments, which then consisted of two flights of five minutes each and "numerous flights in straight lines, in circles, and over S-shaped courses, in calms and in winds." The letter added that "we shall be pleased to take up the matter either on a basis of providing machines of agreed specifications, at a contract price, or of furnishing all the scientific and practical information we have accumulated in these years of experimenting together with a license to use our patents, thus putting the government in a position to operate on its own account."

The letter was passed from Congressmen Nevin to the Board of Ordnance and Fortifications who turned it down, saying that the board did not give financial assistance to inventors. The Wrights tried again in October, 1905, but were turned down again because "... the board does not care to formulate any requirements for the performance of a flying machine or take any further action until a machine is produced which by actual operation is shown to be able to produce horizontal flight and to carry an operator." Thus, the ground-bound Army fluffed off the greatest invention of all time without even bothering to investigate the many reports and newspaper accounts of the Wright flights at Dayton during 1904 and 1905.

The Wrights gave an option to a French syndicate in late 1905, but the contract was not consummated. They spent 1906 perfecting their machine and its engine. Meanwhile, knowledge of their rebuffs by the War Department had spread to the Aero Club of America, whose prominent members brought their invention to the attention of President Theodore Roosevelt. The President ordered the Army to take another look, but in a letter dated October 5, 1907, the board wrote the Wrights that no funds were available and that the matter would require congressional action. A personal appeal by Wilbur Wright to the board on December 5, 1907, however, was so convincing that the board members voted to recommend that the Chief Signal Officer draft a set of specifications for a military flying machine.

When the bids were advertised, forty-one proposals were received from inventors, including many nuts who put forth fantastic claims. Nineteen proposals were so ridiculous that they were immediately discarded. One of them was from a prisoner serving time in a Federal prison who offered to build the Army's machine provided he could be released from solitary confinement.

The prices of the remaining twenty-two bids ranged from a low of $500 to a high of $1 million. Only three of these twenty-two bids complied with the stipulations of Specification No. 486: A. M. Herring of New York City, $20,000 and 180 days; J. F. Scott of Chicago, $1,000 and 185 days; and the Wright brothers, $25,000 and 200 days. However, first Scott withdrew his bid, and then Herring, leaving the Wrights as the sole contenders. The acceptance of their bid on February 10, 1908, had brought Orville Wright to Fort Myer in August.

Orville Wright began his preliminary test flights on September 3, marking the first time I ever saw an airplane fly. The flight lasted only one minute, eleven seconds, but gave me a chance to see how it was done. The *Flyer* was placed on the starting track and launching was accomplished by resting the skids on a small two-wheeled truck placed with its tandem wheels on a monorail track. The weight was lifted to the top of the starting tower by a system of ropes passed through a pulley block fastened near the outer end of the monorail track, thence back to the airplane, and to a trigger which was placed within easy reach of the pilot. The weight was lifted to the top of the starting tower. The engine was then started, and when Orville felt that it sounded just right to him, would give a nod to the crewmen, and then trigger the starting device. The weight would come down to earth with a loud thump, and the plane would catapult down the rail. After a run of about 50 feet, it would leave the track and head down the parade ground.

Between September 3 and 8 Orville made five test flights for a total of twenty-five minutes, five seconds flying time. All flights were made straight ahead without turns so he could make minor adjustments and become accustomed to the very small open space which was surrounded by tall trees and barracks.

On the eighth, Tom Selfridge and I received orders to proceed to St. Joseph, Missouri, along with eight enlisted men, *Dirigible No. 1*, and its equipment. We were to exhibit the Army's new "weapon" at the Missouri State Fair, and make exhibition flights. The men and I left immediately so we could get set up in St. Joe and make such test flights as were considered necessary prior to the opening of the fair on September 21. Selfridge remained behind temporarily to participate in the Wright tests.

While I was en route to Missouri Orville broke the former 1905 endurance record of thirty-eight minutes, and established a new one of fifty-seven minutes, twenty-five seconds, on the morning of September 9, and still another one of one hour, two minutes, fifteen seconds, that afternoon. Encouraged by this, Orville then allowed Frank Lahm to accompany him for a few minutes on one flight, and Lahm thus became the first military passenger in the history of United States military aviation.

Orville kept on setting world endurance records. On September 10 he flew solo for one hour, five minutes, then one hour, ten minutes, the next day. He then demonstrated the *Flyer's* maneuverability by doing figure eights for the crowds that gathered to watch this new wonder of the world. On the twelfth he carried Maj. George O. Squier, president of the board, for a new passenger-carrying world record of nine minutes, six seconds, and then set a new solo record of one hour, fifteen minutes, by circling the parade ground seventy-one times at 300 feet.

As news of each new record spread, the crowds making the trip to Fort Myer each day got larger. There had been many skeptics about the claims of the Wrights before the Myer trials began. The *New York Globe* had editorialized:

> Nothing in any way approaching such a machine has ever been constructed (the Wright brothers' claims still await public confirmation), and the man who has achieved such a success would have, or at least should have, no need of competing in a contest where the successful bidder might be given his trial because his offer was a few hundred or a thousand dollars lower than that of someone else.

But what hundreds had seen almost daily at Myer could not be denied. The spectators soon learned, however, that Orville Wright was a perfectionist, and refused to take risks with his valuable invention. If the wind was too high, or from the wrong direction, he wouldn't fly. He fussed and fumed over his engine and its controls, much like a trapeze artist at a circus. He was determined to fulfill his contractual obligations, and refused to let the impatience of the crowds or the frequent visits of President Taft and the Washington social set persuade him to risk his machine if he didn't feel the conditions were exactly right.

In spite of the extreme care he took with himself and the *Flyer,* tragedy was to mar the tests and almost doom public confidence in the heavier-than-air machine. The final test flight was scheduled for September 17 in the late afternoon. Two thousand people had crowded around the parade ground as a result of increased press coverage which had made the Wrights front-page copy. The *Evening Star* had reported that "Selfridge will be the lucky man tomorrow to fly with Wright as the passenger."

I'm sure that Orville was full of confidence that day, but I'm not so sure about Selfridge. A slight, quiet man who seldom smiled, I recall him as being tense and nervous about flying in our balloon tests, although he certainly never shirked or tried to avoid his turn in the air. He had the attitude of a professional soldier who did as he was ordered. Although Selfridge is said to have volunteered to be the official observer on this final test flight, I am not so sure.

After Orville had checked the machine over thoroughly, he told Selfridge to get in. The crowd was moved back by the mounted cavalry patrol, and Orville took his seat and started the engine. At exactly 5:14 P.M., the weights fell down from the tower, and the *Flyer* skittered down the launching track. It hung close to the ground for a while, and then clawed slowly into the sky to miss the artillery sheds at the other end of the parade ground.

Orville circled the field slowly three times at about 150 feet, but the angle of bank seemed steeper than anyone had seen before. On the fourth circle, as the *Flyer* was heading down wind, there was a puff of smoke, and the plane nose-dived to the ground like a duck shot on the wing. The graceful, delicate machine was suddenly nothing but a smoking jumble of muslin, wood, and wires.

The crowd rushed forward, and soldiers immediately began

pulling at the wreckage to get at Orville and Selfridge. Orville was reached first and gently lifted out. He had a broken thigh, several broken ribs, and lacerations about the head and hands. Dazed, but conscious, he was rushed to the post hospital nearby.

While the soldiers were trying to extricate Selfridge, one of them called to him and asked if he was all right. All Tom could say was, "Take this damn thing off my back." They were the last words he ever uttered. He lapsed into unconsciousness, and died three hours later in Walter Reed Hospital. He thus became one of the first to fly—and the first to die.

While the death of Tom Selfridge was a personal tragedy to all of us who knew him, Maj. George O. Squier told the press that the fatal accident would be only a temporary setback:

> Of course we deplore the accident, but no one who saw the flights of the last four days at Fort Myer could doubt for an instant that the problem of aerial navigation was solved.
>
> If Mr. Wright should never again enter an airplane, his work last week at Fort Myer will have secured him a lasting place in history as the man who showed the world that mechanical flight was an assured success. No one seems to realize at this close range what a revolution the flights portend. The problem is solved, and it remains only to work out the details.

Squier, as head of the Aeronautical Board, ordered that the exact point of the *Flyer's* crash be ascertained "to mark definitely where Selfridge fell in order that at some subsequent time a monument may be erected in his memory." In 1957 the House of Representatives passed a bill authorizing a memorial to Selfridge to be erected on or near the spot where he ended his last flight. Selfridge Air Force Base, Michigan, is named for him. Since then it has been customary to name Air Force bases after deceased fliers.

The cause of this first fatal accident in the history of American heavier-than-air-flight was propeller failure. The propellers, installed only that day, were new ones—eight inches longer than any used previously. A crack developed in one blade of the right prop, and a severe vibration developed which caused the blade to strike a stay wire connected to the tail assembly. Wind pressure on the tail caused the plane to nose dive into the ground from about 100 feet.

Orville Wright remained in the hospital for about seven weeks before he was released to return to Dayton. Before he left, however, he was notified that he would be given an extension of the contract completion date until the summer of 1909. Orville assured the War Department that the accident in no way had caused him to lose confidence in their invention. “I'll be back next year with a new machine,” he told Squier. “We will make good on our contract.”

On the day that Selfridge was killed, I was getting *Dirigible No. 1* ready for its first flight and waiting for Tom Selfridge to arrive in St. Joseph. The telegram informing me of his death was a deep shock because I guess I didn't realize that people could be killed in a flying machine. My first reaction was to call off the dirigible exhibition, but I recalled Tom Baldwin's stories about his experience as a circus performer and how he said it was a tradition in show business that “the show must go on,” no matter what might happen to any individual performer. Therefore, with Sgt. Ed Ward as my rear pilot, we flew as scheduled on September 18 to please the bloodthirsty crowd that came to see two “crazy aeronauts” kill themselves. I was just as determined that I would give them a good show, but I checked all our equipment thoroughly to make sure it would not be a Roman holiday which would end up with our blood spilled.

Frank Lahm was named as the replacement for Selfridge, and he arrived on the twenty-first. The two of us made eight flights during the next four days for a total time of about one hour in the air. At the end of the fair, *Dirigible No. 1* was deflated and shipped to Fort Myer where it was stored until the following spring. I went back to my desk in the office of the Chief Signal Officer, shuffling papers.

In May 1909, Lahm and I were sent to Fort Omaha, Nebraska,

with *Dirigible No. 1* so that we could train three other officers, Lieuts. John G. Winter, Raymond S. Bamberger, and Oliver A. Dickinson, as pilots. The unpacking and assembling of *No. 1* turned out to be a real chore, and taught us some valuable lessons about how *not* to store gas bags and mechanical parts during a winter. We made our first flight on May 26, 1909, but had to land quickly after only nine minutes of flight because of a broken fuel line. We finally had a successful test hop on the thirty-first, and began training our three students.

Minor engine troubles plagued us every day, and we could never spend much time in the air before some mishap or other would force us back to the ground. We had a major repair job on the gas envelope one day while Lahm and Winter were flying. The wind was light when they first took off, but later got fairly brisk. Lahm and Winter soon found that the wind velocity was stronger than the speed of the dirigible and they could make no headway as they tried to land. Even with the engine wide open, the craft started turning down wind and the ground crew seized the drag rope to keep it from blowing away. The few men who could grab the rope weren't enough, and they were pulled off their feet and dragged through a fence before they had to let go. The propeller struck a telephone pole and the envelope got entangled in the wires. We had to borrow ladders from the post fire marshal to get Lahm and Winters down.

Lahm and I taught the three young lieutenants all we knew about the fine art of motorized lighter-than-air flight by the middle of June, which certainly wasn't very much. The War Department ordered the two of us back to Washington to participate in the Wright tests. I had no regrets bidding *Dirigible No. 1* adieu because the little time I had in the air trying to manage that flimsy gas sausage convinced me that dirigibles didn't have much of a future. Although the history books say I was eligible to get F.A.I. Dirigible Balloon Pilot Certificate No. 1, it didn't interest me at all. Lahm got Certificate No.2 but he was a qualified balloon pilot and looked at the honor a little differently. I never flew a dirigible again. Flying around the sky suspended beneath a bag full of highly inflammable gas and being at the mercy of the winds did not show any military potential to me. It has always mystified me why the Navy held on to their gas bags so long, especially when they saw what happened to the German Zeppelins in World War I over England.

I guess my feelings about the old bags got around the War Department, because when I got back to Washington I was asked by the General Staff to give my opinions about them. Not being used to beating around the bush, I gave my opinions, unaware that they were counter to the thinking of the Chief Signal Officer and his non-flying but nevertheless expert staff. Based on my study at Leavenworth in 1907, my many talks with Tom Baldwin, and my reading of intelligence reports on the military use of dirigibles in England, France, and Germany, I recommended that no more dirigibles be procured, and "that we concentrate our interest in development of heavier-than-air vehicles." The Wright machine had already flown longer, faster, and more reliably than *Dirigible No. 1*. I could not see what anyone could ever do with an ungainly bag of gas to make it go much faster.

My views did not go unheeded. They were so unpopular that I was immediately earmarked by my superiors as a dangerous, erratic nut who would have to be watched carefully lest his radical thoughts get to the public.

Chapter Five

A Day to Be Remembered

When June 1909 rolled around, the Wrights had firmly established the fact in the public mind that their flying machine could indeed fly. On June 9 President Taft invited them to the White House, and presented them with the Aero Club Medal in long-delayed recognition of their achievements. Eight days later a celebration in their honor was given in their home town of Dayton, Ohio, where they received medals from the city and state and a special medal authorized by the Congress which was presented by General Allen.

Orville and Wilbur returned to Washington on June 20, intent upon completing the trials that had been interrupted by the tragic accident the previous September. The machine they brought this time was an improved version, which incorporated a number of engineering changes. Structural changes had been made to prevent brace wires from fouling the propeller, the method of attaching the rear rudder had been changed by using a solid wire in place of an elastic stay, and the height of the skids had been increased in front, making them resemble runners on a sleigh. Between the control surfaces on the front elevator two semicircular vertical fins were installed to prevent spinning. Two control levers—one to the right, and the other to the left of the two pilots' seats—controlled the elevator, as in the 1908 machine, but the rudder in the rear was operated by one of the two parts of a split-hinged lever located between the two seats; the other part operated the wing warping for lateral balance. The combined action of these two parts caused the machine to bank at the correct angle. There were no instruments, but the Wrights had fastened a piece of string about eight inches long with a weight attached to the horizontal crossbar between the front ends of the two skids. This served as a turn and bank indicator, and showed by its angle whether the airplane was sliding inward or skidding outward during a banking turn. If the string inclined toward the pilot, it indicated a loss of forward speed and danger of a stall.

The engine, which was practically the same as that in the 1908 plane, was controlled by a foot pedal placed on the footrest crossbar out in front, midway between the two seats. The pedal was connected to the magneto and served to advance or retard the spark. In order to start the engine, it was necessary for someone on the ground to hold a piece of waste, saturated with gasoline, over an open tube into which air was sucked as the engine was turned over by cranking both propellers. The gasoline tank held enough fuel (thirteen gallons) for a flight of three and one-half hours.

The launching apparatus remained the same as before, with a monorail and weight tower. On windy days the airplane was headed into the wind, balanced on the monorail, and set in motion by the propellers without the use of the catapult. On calm days the catapult was used. The skids rested on a small car with two grooved wheels set tandem on the rail. A rope, hooked to the car, ran through a pulley at the front end of the starting track, under the track and through a pulley at the other end; it then passed through a third pulley at the top of a 30-foot pyramid of braced posts, which stood about 10 feet behind the monorail. Weights, totaling almost 2,000 pounds, were attached to the end of the rope which hung down inside the pyramid. The plane was started by drawing it back to the beginning of the track, the weights were hoisted by a block and tackle, the propellers were started, and the weights were allowed to drop. The force of the falling weight, plus the power of the propellers, pulled the plane rapidly along the track. When flying speed was reached, the pilot pulled back on the elevator lever and the plane soared into the air.

It took a month for the Wrights to assemble the machine and test fly it until they were satisfied with its performance. As before, I hung around them everyday and pestered them with questions. They tinkered and fussed and muttered to themselves from dawn to dusk, and it seemed as if they would never say they were ready to go. There were many delays of their practice flights because of high winds or rain showers, and those of us on the Aeronautical Board were kidded unmercifully for being "birdmen sitting on the wall," which we seemed to be doing a lot of.

To while away the time, I read a book about flying, written by an Englishman named Lancaster. The questions I asked the Wrights were based on what I had read. Finally, Wilbur stopped working after I had put a question to him, and asked, "What in the world have you been reading to ask something like that?"

"Everything I can lay my hands on," I replied.

Obviously exasperated, Wilbur said, "There are no books worth reading on the subject of flying. You get your hands on that machine over there if you really want to learn about it."

This was a rare invitation for the Wrights to extend to anyone, and I was delighted. I donned my coveralls, stuck a pair of pliers, a screw driver, cotton waste, and a bar of soap in my pockets as I had done before, and got to work.

Close-mouthed and quiet, the only thing the Wrights liked to discuss was their machine and the forthcoming tests. They had no interest in socializing in Washington, and as far as I know never accepted a single invitation during the test period. They granted no interviews with the press, and turned all requests away.

Orville was the more talkative of the two, which wasn't saying much. He was amiable and kind-faced. Wilbur, prematurely bald, about forty then, had deeper furrows in his face and seemed to look at you with a kind of reserved suspicion. When you spoke to the two of them, it would be Orville who would answer, and Wilbur would either nod his assent or add an incomplete sentence as his way of corroborating what his younger brother had said. At no time did I ever hear either of them ever render a hasty or ill-considered answer to any question I asked, and sometimes they would take so long to reply that I wondered if they had heard me.

Apparently, the brothers Wright had decided as a matter of family honor that Orville would fly all the tests at Myer, because I don't recall Wilbur ever flying the new model until after the tests were completed. After a number of short flights, Orville made an endurance flight of one hour, twenty minutes, on July 20. On the twenty-seventh he announced that he was ready to resume the tests where he had left off the year before. With Frank Lahm as passenger, he flew one hour, twelve minutes, and forty seconds, thus more than fulfilling the requirement to remain in the air for an hour with a passenger. It established a new record for a two-man flight.

Next and last specification to be satisfied was the speed test over a measured five-mile course. Because of my previous map-making experience, Major Squier asked me to layout the course. I chose a rise of ground at Alexandria, Virginia, almost due south of Fort Myer, called Shooter's Hill, to be used as a turning point. It is the present site of the George Washington Masonic Memorial, the cornerstone of which had been laid not long before. I chose this site because it was above the surrounding terrain and should have been easy to navigate toward at an altitude of 100 or so feet. I arranged for a temporary telephone as well as a telegraph line between the two points, the latter to be used for sending a signal at the exact second the plane crossed the measured mark at Alexandria. The telegraph instrument was actually placed on the Masonic Memorial cornerstone.

Remembering my dirigible experience, I was concerned that we would wander off our course from Myer to Alexandria, so I arranged for a small captive balloon to be ascended from Shooter's Hill and one anchored about halfway between the two points. Finally, it was announced on July 29 that the cross-country speed test would be attempted the next day. The only question among the board members was who was to be the navigator-passenger on this all-important flight. It was decided that the Wrights should make their choice. They did: me.

I would like to think that I was chosen on the basis of intellectual and technical ability, but I found out later that it was my short stature, light weight, and map-reading experience that had tipped the decision in my favor. A heavier man would have added weight to the plane which would have slowed it down. A man who couldn't read a map and had not been aloft before might get the pilot lost, add unnecessary distance to the flight, and thus decrease the speed average. A 10 per cent bonus was riding on every mile they could squeeze past 40 miles an hour.

About 7,000 people showed up on the afternoon of July 30, which was less than usual. Rain showers discouraged many from making the trip because they knew that the Wrights never flew if the weather was unfavorable. About four o'clock it looked as though the sky was clearing and the wind was dying down. Orville told the board members he would be ready in about one hour and a half. Maj. Charles McK. Saltzman and Lieut. George C. Sweet, a Navy observer, quickly drove to Shooter's Hill with the field telephone to let us know when everything was ready there. The Wrights pushed their machine to the starting rail and made many adjustments to the engine and guy wires. I put two stop watches around my neck and got into the passenger seat. I strapped a box compass to my left thigh, lashed an aneroid barometer to my right thigh, and jammed a map into my belt.*

Orville warmed up the engine until he was satisfied with it, and climbed aboard. "If I have any trouble," he shouted above the roar of the engine, "I'll land in a field or the thickest clump of trees I can find."

I nodded and gulped. I had picked a course with no fields of any kind en route. It was too late to do anything about it now, so I grabbed the edge of the seat with both hands and waited. Orville revved up the engine, released the trigger, and the machine started down the rail. We skimmed over the grass for a few feet to gain speed, and then climbed for altitude. As we started to circle, Wilbur ran to the center of the field below us with a stop watch in one hand and a signal flag in the other. We made two complete circles of the field, gaining altitude (125 feet), and then Orville swung sharply over the starting line. I flicked one stop watch and pointed out the exact course we should follow to Shooter's Hill.

* These were the only instruments I ever took aloft with me until 1914.

All twenty-five horses in the engine were functioning perfectly as we skimmed over the treetops toward the first balloon. The air was bumpy, and I had the feeling that there were moments when Orville didn't have full control of the machine as we dipped groundward. It was as if someone on the ground had a string attached to us and would pull it occasionally as they would a kite. But each time Orville would raise the elevators slightly, and we would gain back the lost altitude.

We reached Shooter's Hill all right, and I flicked the second stop watch. There was a crowd on the brow of the hill, and I could see them wave their umbrellas and handkerchiefs. It seemed to me that the angle of bank of the plane was awfully steep as we rounded the turn and the wing tip was much too close to the tops of the trees. A down draft hit us, and I thought we were going to cartwheel into them for sure. We straightened out, however, and started back for Myer. Going down wind now, our ground speed increased and Orville climbed until we reached 400 feet—a world's altitude record. As we neared Myer, Orville nosed down to pick up speed, and aimed at the starting tower. I flicked the stop watch off as we crossed the starting line and relaxed as he made a circle over Arlington Cemetery, cut off the engine, and glided in for a fairly smooth landing amid a cloud of dust.

Wilbur rushed up to us, and it was the first time I ever saw him with a smile on his face. I learned later that he had experienced some excruciating moments of doubt when we had disappeared below the level of the trees around the parade ground on our outward trip.

The crowd was larger when we landed than when we had departed. President Taft had not been present for the takeoff but had seen us land, and sent a messenger through the crowds to us with a note of congratulation. I had the feeling, though, that some of the citizens present were disappointed. From my experience with the crowds that had witnessed the previous test flights, I had no doubt that many of them were disappointed that we had not landed in Arlington Cemetery, and thus provided them with a real old-fashioned Roman holiday with all the bloody trimmings.

As soon as the board members could get together, we compared our stop watches and determined that the official speed to Alexandria had been 37.735 miles per hour; on the return trip it was calculated at 47.431 miles per hour, with the average officially computed at 42.583 miles per hour. Major Squier asked the Wrights if they wanted to make another trial since the specification allowed them three chances. They replied that they would stand on this their first cross-country flight.

Since they had flown more than 42 miles per hour, it was thought that they would be eligible to collect a bonus of $7,500; however, some government lawyer disallowed the fraction above the 42 miles per hour and they qualified for only $5,000 above their bid price of $25,000. On August 2, 1909, *Aeroplane No. 1* was officially accepted into the inventory of the United States Army. Just as important was the fact that we had established three world records for the United States that date: one for a two-man flight at 42.583 miles per hour; another for a cross-country flight of 10 miles; and a third for an altitude record of 400 feet which I verified with the reading of the aneroid barometer strapped to my leg. July 30, 1909, was truly a significant day in American aviation.

There remained only one final condition to the contract after the final test, and that was the provision for the training of two pilots. The commanding officer at Fort Myer by this time was completely fed-up with the flight tests because they had tied up his drill field and interfered with his summer training schedule. This didn't disturb the Wrights because the Myer drill ground was too small for safety anyway, and the proximity to trees, buildings, and crowds bothered them. As a result, Frank Lahm was detailed to find a better location.

Lahm had some ideas for better fields from his free-balloon

ascensions, and looked them over on horseback during the month of August. He chose a field near the Maryland Agricultural College at College Park (now the University of Maryland), and both the Wrights and the Aeronautical Board approved. The field was leased and promptly cleared of trees and brush by Army troops. A small hangar was erected and the monorail starting track laid down.

General Allen designated Frank Lahm and me to be the officers who would get checked out as the first two Army pilots. Meanwhile, Wilbur Wright had taken another plane to New York, and gave the big city its first view of flying machines. His first flight on September 29 was from the Governors bland drill field on a course around the Statue of Liberty and return. Then, on October 4, he flew up the Hudson River to Grant's tomb and back. The next day he returned to Washington, ready to complete his government contract and teach us to fly.

I was happy to have been selected as one of the first two Army pilots, but my joy was short-lived. Without any advance warning I was handed orders to attend the International Congress of Aeronautics at Nancy, France, as the official delegate from the United States. I was then to go to Frankfurt, Germany, to visit a German aeronautical exhibition, thence back to Paris to report to a Major Mott, the military attaché there, and tell him what I had seen.

I was bitter about this sudden change in plans. My greatest desire was to be one of the first two pilots to solo in the Army's first heavier-than-air machine. However, having had a decade of experience in the Army's University of Hard Knocks, I kept my mouth shut, and packed my bags. Lieut. Frederic E. Humphreys, an officer from the Corps of Engineers, was appointed to fill my spot.

My visit to Nancy was interesting because I had never been to France before, and I was amazed at what the French had done on their own to try to get ahead of the Wrights. I detected a “Yankee, go home” attitude, but I attributed it to a national jealousy at the success Wilbur Wright had achieved there a few months before.

Neither the trip to Nancy nor the trip to Frankfurt provided any useful information for me. By the time I got to the exhibition hall in Frankfurt, most of the exhibits had been removed, and my late arrival did not cheer any of the Germans I met into giving me any information. By the time I got to Paris to report to Major Matt, and found that all information of any intelligence value from both exhibits had already been reported, I was furious. To add insult to injury, I wasn't allowed to see the reports because I had no orders which said I had a “need to know” such classified information. Both Matt and I agreed that a lot of taxpayers' money had been spent to give me an enforced tour of Europe.

I thought someone had really goofed in selecting me for the European trip until shortly after I returned to Washington in late October. There, over a little liquid refreshment, Maj. Charles McK. Saltzman (later Chief Signal Officer) told me that the trip was actually a reprimand for the opinions and recommendations I had rendered about the future of the dirigible. I had gone 180 degrees away from the prevailing opinion among the staff members of the Signal Office. They had observed how elated I was at being chosen for the heavier-than-air training, and decided to teach me a lesson. Anyone, especially a junior officer who went against staff thinking on any subject in those days, invited a reprimand for himself rather than a reward for daring to think imaginatively. The staff had to send someone to Europe, they thought, so they drafted orders for General Allen's signature. I was, they told the general, the most qualified man to go since I had just completed the tests with the Wrights.

“You'd better watch your step from here on, Foulois,” Saltzman told me, “or they'll have you stringing telephone wires up in the Klondike.” I got the message, but it didn't change my mind about dirigibles.

I returned to the States on October 19, and found that Lahm and Humphreys had started their instruction with Wilbur

Wright, as planned, at College Park. They had each had a few minutes' flying time, when Wilbur decided to try an experiment. He wanted to see whether his machine could take off from the monorail track under its own engine power and without the thrust of the falling weight. He found that he could, provided the plane carried only one person and the monorail was pointed into the wind. This was what we would call now a technological breakthrough, and proved a point: the flying machine had great potential, but its potential would be directly proportional to the power of the engine that propelled it. The Wrights recognized this, and were constantly striving to make the engine more efficient so it could deliver more power to enable it to go faster, higher, and farther. This, after all, is what aviation is all about.

My first flight with Wilbur took place on October 23. On the twenty-sixth Humphreys made his first solo flight, which puts him down as the first official military pilot in American history. He was followed a few minutes later by Frank Lahm. They followed these initial solos with additional flights to give them skill and confidence.

Humphreys took me with him for a ten-minute flight on the thirtieth, and this was another aviation first. It was the first time that a military man had taken up another as a passenger. Later that day we went up together for a thirty-nine-minute flight, which was terminated by the breakage of a gear tooth in the engine magneto. It was another pair of military aviation firsts—the first engine failure in flight and the first successful "dead-stick" landing.*

On November 3 the Army chalked up another first for the United States when Frank Lahm carried his first passenger, Lieut. George C. Sweet, U.S.N. Thus Sweet became the first United States naval officer to fly in a heavier-than-air machine.**

* The accident that had killed Tom Selfridge was owing to propeller failure and not engine failure.

** Wilbur Wright gave a ride to Mrs. Ralph H. Van Deman, wife of Army Captain Van Deman on the morning of October 27, 1909. A close friend of Miss Katherine Wright, Wilbur's sister, Mrs. Van Deman was the first woman in America to be carried as a passenger in an airplane.

The morning of November 5 marked the end of heavier-than-air training for the Army in 1909. On that day Lahm and Humphreys were flying together, when, with Lahm piloting, they made a turn at a low altitude and caught the left wing tip on the ground, causing the plane to cartwheel into the ground. The plane's skids broke, and the right wing was badly bent. Fortunately, neither of the pilots was injured. They had just achieved another pair of firsts—the first time two qualified Army pilots had flown together and the first time an Army crew had cracked up an Army plane.

Wilber could not repair *No. 1* with the spare parts he had on hand and had to order replacement parts from Dayton. While we waited at College Park, the weather turned cold and strong autumn winds swept across the field from the northwest. "Even if we had an airplane ready to fly," Wilbur told us, "we couldn't fly in winds like this."

Wilbur was right, of course. Besides the hazardous cross winds which could make student training extremely dangerous there was the biting cold. There was no windscreen or enclosed cockpit on *No. 1*, and we had no warm flying clothes such as we developed later. If we had flown at all, we would have had to wear our long, bulky Army overcoats, riding boots, and thick gloves. Sitting on the leading edge of the wing with our feet on the rudder bar without adequate clothing that would permit freedom of movement, we would probably have been the first fliers in history to have frozen to death in flight—an honor we did not care to have.

It was inevitable that someone would recommend that flying training be moved to a warmer climate. In late November General Allen ordered that operations cease at College Park and move to Fort Sam Houston, Texas. Arrangements were made for the shipment of *No. 1* as soon as it was repaired and the transfer of the enlisted mechanics. There was a hitch in the plans, however, when it came to the officers. Humphreys had been assigned to the College Park operation only as a temporary replacement for me so that Wilbur could satisfy the requirement to train two pilots. The chief of the Corps of Engineers ordered him back to his engineering duties after he had cracked up with Lahm. At the same time, Lahm, a cavalry officer, was ordered to return to his regiment because of a statute we called the "Manchu Law" which required all line officers to return to duty with troops in their own basic branch of the Army after detached service of four years. In November, 1909, the entire combat air arm of the United States, therefore, consisted of one badly damaged airplane, eight attached enlisted mechanics, one civilian mechanic, and one untrained pilot—me!*

Since the Wrights had completed all the requirements of the contract when Humphreys and Lahm soloed, Wilbur was paid after the November 5 crack-up. He had no further legal obligations, but he insisted that the government would not be charged for the repairs to *No. 1*, and no bill was ever rendered. He supervised the repair of *No. 1*, and departed as we got it ready for shipment. Before he left, however, he told me to write to him if I had any questions about the airplane or flying, and he would do his best to answer them. I appreciated this gesture, but the full force of his offer did not really hit me until he was gone. I had not soloed yet, and I wondered if the government would consider hiring the Wrights as instructors, or allow Humphreys and Lahm to return to help me out. I soon found that no money was available for the first idea, and War Department policy could not be bent a little for the second. Besides, I was told, flying wasn't all that difficult, and I had already had several flights—fifty-four minutes of instruction time with Wilbur Wright and about two hours as passenger with Lieutenant Humphreys.

In December 1909, I was called into General Allen's office in Washington and officially told to take *Aeroplane No. 1* to Fort Sam Houston. It was then he gave me his famous order to

* Humphreys was so disenchanted with not being able to continue flying after being the first to solo, he resigned from the Army a few months later. He returned to flying briefly during World War I and afterward became a brigadier general in the New York National Guard as commander of an Engineer regiment.

"take plenty of spare parts, and teach yourself to fly." While I was packing I got an amendment to this order which told me to proceed by way of Chicago, Illinois, where I was to put *No. 1* on display at the Coliseum during the *Fifth Annual Electrical Show* of the Chicago Electrical Trade Association for the two-week period, January 15 to 29, 1910.

The *Electrical Show* was a big deal in those days, and the newspapers played it up every way they could think of. This year the show was to be "the largest ever," where exhibits would include "everything that's new in light, heat, and power for the home, office, store, factory, or farm." The papers called the Coliseum a "fairyland." Its roof was draped with 100,000 yards of hanging gold and silver tinsel; six powerful searchlights with color slides in front of the lenses gave an ever-changing rainbow effect. There were more than 150 exhibitors showing off their wares, and in the center of all this we were asked to hoist *No. 1* up in the air where it could be seen from anywhere on the floor. It was billed as the main attraction.

There was a method to our madness about putting the plane on display at an *Electrical Show*. I believed what I had written in my thesis about the possibility of the wireless being used in an airplane, It had never been tried before, and it seemed to me that this exhibition would be the ideal opportunity to try it out.

My men and I assembled *No. 1* and hoisted it into position. After it was in place, the newspaper reporters had their first headlines and first hard news of the show:

AVIATOR ESCAPES BAD FALL
Lieutenant Foulois Grabs Beam
When Ladder Leading to Aeroplane
Gives Way

> The Foulois airplane was securely hung about twenty-five feet above the floor yesterday by men of the United States signal service, and after work was completed, Lieut. Benjamin D. Foulois narrowly missed his first accident since he has followed

> the hazardous profession of imitating birds. The little army officer has made forty successful flights, one of which was the famous trip with Wilbur Wright from Washington to Alexandria, Virginia, and return last July.
>
> When the machine had been safely "anchored" yesterday, Lieutenant Foulois started to ascend the ladder leading to it. He had reached one of the top rounds, when a swerve of the machine allowed the ladder to swing free, and down it crashed to the floor. In most instances a man at the top of the ladder would have been dashed to the floor and unquestionably severely injured or killed, but the lieutenant quickly grabbed a support with one hand the instant he felt the ladder give, and saved himself from the fall.

I won't comment on the exaggerations or the errors in this reporter's story, but I had long before learned to have an alternate plan when climbing ladders which I had learned working with my father. But the main effect of the news story was to stir up interest in the show and its main attraction, the Wright airplane.

I made friends with Frank L. Perry, a ham radio operator from Chicago, and we decided to see if we could rig up a wireless telegraph and actually send messages between a ground station and the airplane. We strung two antennas on the craft, one on the upper wing and the other along the skids or "runners" as we called them. We took a standard Army telegraph sending and receiving set and installed it behind the pilot's seat along with a set of batteries. Perry rigged up a similar station in a far corner of the Coliseum and we both crossed our fingers.

I know the elation that Alexander Graham Bell must have felt when I received Perry's first message while I was perched up in the plane and I sent back an answer. It was another aviation "first" and immediately got more headlines. This story, which appeared in the *Chicago-American* on January 18, 1910, was typical:

USES WIRELESS
ON AIRSHIP
FOR FIRST TIME

> For the first time in the history of aeronautics and wireless telegraphy, an aerogram was sent to an airship yesterday. It happened at the *Electrical Show* at the Coliseum and by a coincidence the achievement which advances aeroplaning and wireless each a step occurred on the one hundred and fourth birthday anniversary of Benjamin Franklin, who startled the scientific world by drawing electricity from the storm clouds with a kite and a key.
>
> Franklin, as the most extravagant of dreamers, could never have foreseen the day when it would be easily possible and practicable to send a message to the clouds without wires...

Encouraged by this success, my men and I decided to put the icing on the cake in so far as the show's main attraction was concerned. Billed as the star performer, we thought *No. 1* would appear more realistic if we rigged up an electric battery drive for the engine, sufficient to operate the propellers at about 400 revolutions per minute. It was arranged that on opening night the bands would play, a circus barker would give a stirring introduction, and at the magic moment the cymbals would clash, the searchlights would be turned on the airplane, and one of my men would throw the switch to start the propellers. This was done, and the applause was deafening. The star attraction had done its part well.

The second act on the program was Miss Blanche Mehaffey, a popular prima donna of those days. Miss Mehaffey made only one mistake. She appeared on a small balcony which happened to be directly in the air stream from the propellers of *No. 1*. When she opened up her tonsils with the hit song of the day "I've Got Rings on My Fingers," the blast from the props blew most of the song down her throat and her dress up around her necklace. She became a few degrees more than furious and refused to sing another note until we stopped the propellers.

The newspapers had another headline for the next day's programs: WIRELESS STOPS AEROPLANE TO PLEASE SINGER. What prompted it was that I sent the message to my man in the machine by wireless, because, as one newspaperman reported, "When an aeroplane interferes with the singing of a prima donna, cut out the aeroplane, even if it has to be done by wireless." This same reporter gave us another aviation "first" when he ended his article by saying, "History tells that professional singers are easily annoyed, but this is the first time an aeroplane had to be stopped by wireless telegraph to ensure an appearance on the programme.

While the telegraph experiments had been a lot of fun, they had a very serious significance for the future of aviation. I could not conceive of airplanes realizing their maximum potential both for the military and commercial user unless two-way contact could be continuously maintained between planes in the air and between planes and the ground. Our experiments answered an important question which the experts had argued about, and that was how an airplane radio set could be grounded so that the electrical current could flow. The answer was simply to ground it on the air frame. This was as important a technological breakthrough to aviation radio then as the airplane had been to human flight.

My reports on the radio experiments were read by the Signal Corps staff in Washington and caused some red faces among the "experts" who had publicly stated that two-way wireless communication with airplanes was impossible. If it is possible, some said, it was certainly not practicable because, as one Signal Corps officer explained, "there would have to be radio stations all over the country tuned in just waiting for the aeroplane drivers to say something important enough to be worth the expense of keeping the radio operators so employed." I have often wondered where this fellow ended up.

When we concluded our business in Chicago, we loaded *No. 1* on the train and accompanied it to Fort Sam Houston in San Antonio. During the month of February 1910, we erected a small hangar on the mounted drill ground located in the northwest section of the reservation. *No. 1* was given a thorough overhaul as it was reassembled. On March 2 I made my first solo, landing, takeoff, and crash. This group of personal "firsts" covered in the first chapter, were also aviation firsts for the state of Texas since up to that time no one had yet ventured to that part of the country to show off the airplane.

My first crash put *No. 1* in the shop for a week. On March 12, I made five flights and learned about gusty Texas winds. On the last of these flights, I cut the engine off for the landing when a sharp down draft hit the plane and literally took the controls out of my hands. As the plane went down, I went up. My head hit the top wing and I received a sudden headache. As I came down, the plane came back up and I was back in my seat with a whoomp that knocked the breath out of me. The only reason I wasn't thrown out of the plane was that there were two truss wires in front of the pilot's seat which prevented me from going overboard. With throbbing head and aching seat, I jammed the nose of the plane downward and managed to regain control of my bucking bronco before it stalled. The landing was as near to a crash without being one as could be imagined.

Wilbur Wright had written me one of his basic rules of flying: "When in trouble in the air, put the nose down—and stick to the ship." What he meant was that the first reaction of a pilot must be to prevent a murderous stall by pushing the nose down, and that a pilot should not abandon his ship because that first step out of a flying airplane could ruin his whole day, especially since parachutes had not yet been invented. While I had remembered the first part of the rule, I realized that there wasn't much I could have done about staying in my seat voluntarily in rough air. So the next day, after my headache subsided, I went to the saddlery shop of the field artillery unit at Fort Sam and got a saddle strap with which I lashed myself to the pilot's seat on subsequent flights. This strap was the world's first aircraft safety belt.

The Field Artillery unit was commanded by Capt. Fred

Austin, who later became Chief of Field Artillery. Austin and I became good friends and he was very interested in my efforts to teach myself to fly. He made it a point to have his battery out for drill on the days when I was in the air, "so I'll be able to help the Air Force pick up the pieces," he would say. He wasn't disappointed, for one day I was flying over the drill field when I flew into one of those little dusty whirlwinds that I had seen rising from the field occasionally but had never before encountered in the air. The plane was tossed upward and over on one wing at a very steep angle. Remembering Wilbur Wright's rule and my own experience, I cut the engine and jammed the nose down to prevent stalling. However, there was one small but important fact that I overlooked. There wasn't sufficient space between the airplane and the ground fully to recover control. Suffice it to say that my flight was suddenly terminated amid the nonmusical sounds of ripping linen, splitting wood and tangling wires.

Fortunately, I was able to cut the engine and stay in the seat, thanks to the leather safety strap. Austin's men came galloping over to pick up the pieces and pull *No. 1* off its dazed pilot.

There is an old saying everyone has heard to the effect that a good landing was a landing from which the pilot could walk away. This was not one of those since I couldn't walk away this time. My left leg was pinned down by one of the airplane truss wires and I had a nasty wire burn. After fifty-six years I still carry this brand as a reminder of this flight. It is the only scar, physically speaking, that I ever acquired in my entire active aviation service.

The combination of my limited experience in landing with a dead engine (the engine on *No. 1* was always cut off before landing), the bad bucking habits of *No. 1* in gusty winds, and forced landings because of the erratic temperament of the engine, kept the machine in the shop more days than it was out. General Allen had said "take plenty of spare parts" when he sent me to Texas, but he didn't realize that there were no spare parts. The Army had purchased one airplane, but nowhere in the contract did it say that the contractor had to furnish any extra parts. After my pleas were finally heard along the Potomac, we received our first spare parts (engine cylinders), but not until two months after I needed them. Needless to say, my plumbing experience came in handy trying to make do on the engine until they arrived.

I had been allotted $150 for maintenance purposes, but the repair bills during the first four months of teaching myself to fly exhausted this appropriation. I had to contribute about $300 of my first lieutenant's pay to cover the cost of new materials for wing covering, rudders, elevators, propellers, and runners. These personal expenditures plus much midnight requisitioning in the plumbing, carpentry, paint, blacksmith, and tailor shops at Fort Sam enabled me to carry on a reasonable amount of self-instruction and a fairly steady program of service testing and modification of *No. 1*.

My two major objectives, as I interpreted my mission, were to learn the secrets of flight with a minimum of damage to me and the machine (in that order), and experiment with the airplane design, the engine, and its fuel and oil systems for use in making future airplane and engine specifications. For both of these reasons I kept up a lively correspondence with the Wright brothers. Their advice on several occasions probably saved my life and the airplane as well. I profited by their mistakes and incorporated their design improvements on old *No. 1*. Thus, I was the first, and only, pilot in history to learn to fly by mail. From some of the near misses I had, I guess I was the first person who was literally kept alive by the postman.

While maintenance troubles were continually causing me to lose much valuable flying time, more time was lost on the days I could fly by the necessity for constantly shifting the launching rail because of the ever-shifting Texas wind. In one of my first reports to the Chief Signal Officer I asked that a set of wheels be procured from the Wrights in order to get rid of the cumbersome launching track, starting tower, and landing skids. I received no answer, so I started to design my own wheels. In the midst of my efforts I received a blunt military letter from Washington reminding me that I was first of all a Signal Corps officer and that flying an experimental machine, since it could not be flown very often, was "in addition to your other duties." Thereupon, I was ordered to take charge of the installation of an "electric buzzer annunciator system" on the Army target range at Leon Springs, about 25 miles from San Antonio. So, for the next seventy-five days much of my time and the time of my mechanics was spent plowing and digging trenches, mixing cement, rebuilding target range houses, laying electrical cable, and testing the system so that the range personnel wouldn't have to use semaphore flags to signal scores back and forth.

I learned a year later why I had been sent on this little "additional duty." Seems as though the fact that I had spent my own money to procure parts for *No. 1* had gotten into the local paper and duly noted in Washington. When my request for wheels was received, the Signal Officer's staff decided that it was time to remind a certain lieutenant that he was permitted to fool around with flying machines solely at the win and whim of the Army. This was the second time that I was "disciplined" without knowing it.

As I look back on that "additional duty" at Leon Springs, one thing stands out in my memory, and that was the attitude of my men. On the morning I told them we would have to install the system and still try to get our flying done, they tackled this new task without question, and never once did any of them complain about the extra work or the fact that they were unfamiliar with an electric buzzer annunciator system. In so doing, they laid down the basic foundation for the high standards of morale and duty that have guided the Air Force ever since.

When the job at Leon Springs was finally completed in July 1910, I got back to the construction, installation, and testing of the wheel assembly. O. G. Simmons, our civilian mechanic, and Corp. Glenn R. Madole, had done the preliminary work on it at the San Antonio Arsenal, and after a great deal of trial and error we finally got *No. 1* into the air with its new wheels adapted from a cultivator on August 18, 1910. To our relief, the three-wheel assembly—the Army's first tricycle gear—worked perfectly.

The more I flew *No. 1* in the rough Texas air, the more I worried about the tendency of the old girl to buck like an unbroken cow pony. I got very proficient in taking off and flying in straight and level flight, but every landing broke something. I would fire off a letter to the Wrights, describing what happened, and they would mail back an analysis. Between us, we decided that *No. 1*'s bucking habits in gusty winds, which became a real problem during landing, were owing to the two large elevator surfaces on the front of the air frame. One of these was then placed to the rear, but it only dampened *No. 1*'s bucking habits a little. On September 8 I crashed again, and for the second time Captain Austin's men had to gallop to the rescue and pull the wreckage apart in order to separate *No. 1* from its pilot.

About this time the Wrights wrote me that their *Wright-Type B* plane (*No. 1* was a *Wright-Type A*) had greatly improved flying characteristics. Several of the *Type B*s were going to be flown at an aviation meet at Belmont Park, Long Island, in October, in competition with the Curtiss machine and the leading foreign machines, such as the Farman, the Blériot, and the Antoinette. Since the Wrights had by now completely discarded the use of front elevators I thought this was an important development, and asked permission to attend the meet as an observer. To my surprise, my request was approved.

On one of the first days of the meet, when the 30-mile-per-hour wind kept the other planes on the ground, Walter Brookins, one of the Wright pilots, flew his *Wright-B* in front of the grandstand at about 200 feet. He throttled back and reduced his flying speed to the speed of the wind, and hung there suspended in the air with a zero ground speed. Brookins' fellow pilots and the crowd in the grandstand were astonished. His feat proved a point that had been argued on the ground to the effect that an airplane always had to be going faster than the wind or it would crash.

Another marvelous demonstration of the *Wright-B* was associated with the altitude record competition in which all planes were scheduled to take part. When the time came for the competition, the wind was whipping down the field at 30 miles per hour. The only planes to take off were two *Wright-B*s, piloted by Ralph Johnstone and Arch Hoxey. Another entry, a French Antoinette, tried to get off but piled up at the end of the runway trying to get airborne. All the others remained safely on the ground.

After Johnstone and Hoxey took off, Wilbur Wright tugged my sleeve and motioned me to follow him to the Wright hangar. "This is going to be interesting," he said. "Watch them as they climb. Use the comer of the hangar as a sighting point."

The pair of *Wright-B*s chugged away climbing straight into the wind. When they reached about 5,000 feet, the forward flight of both planes ceased. Both planes, however, continued to climb. Slowly, as I sighted along the hangar wall, both planes were driven backward a distance of five miles to the rear of their starting point but were still climbing. They landed a little later when their gas ran low. Both pilots had broken the world altitude record—flying backward.

To me these two demonstrations meant more than just proving that an airplane *could* fly in a wind greater than the speed of an airplane. The *Wright-B* proved its stability and Wilbur and I agreed on how I could modify old *No. 1* into a reasonable facsimile of the B.

The year 1910 had been one of "unprecedented activity" according to one writer, and I guess it was. The world speed record had been increased 50 per cent to about a mile a minute, and the altitude record was now 8,692 feet. The nonstop flight record was 244 miles, and a plane had stayed aloft five hours and thirty-two minutes. A French pilot had crossed the Alps, and four passengers had been carried from London to Paris. Aircraft engines had increased in power to as much as 100 horsepower. A gun had been fired from an airplane for the first time, and the first bombs were dropped that year.

But while aviation in general seemed to be making progress, the Army seemed to want little of it. In the annual report of the Chief Signal Officer for 1910, General Allen affirmed that the airplane was of interest to the Army mainly for reconnaissance and for carrying messages from one place to another. However, he dosed his report by noting: "The age of mechanical flight is at hand. All first-class powers, except the United States, are providing themselves systematically with aerial fleets." He recommended the purchase of "at least twenty aeroplanes" but stated that "this estimate is considered extremely low."

General Allen's report was duly read by the civilian and military leaders in the War Department, but his recommendations were ignored. It was to be an oft-repeated pattern for the nation's Air Force, which is still being followed in the Space Age.

Chapter Six

Every Flight a 'First'

Throughout the year 1910 the Wright brothers had made many experiments to improve the stability and control of their planes. The result was a complete abandonment of the front elevators in favor of a single flexible surface located near the rudder in the rear. I had discussed these changes with the Wrights, and we worked out the way I should go about incorporating them into old *No. 1*. I returned to Fort Sam and spent the next four months performing a face-lifting operation on the old girl.

While I was trying to get the one-plane Air Force back into the air, Gen. James Allen, Chief Signal Officer, asked the Congress for funds to buy twenty sister ships for *No. 1*. He failed to convince the right people on Capitol Hill. In fact, one congressman, when he was approached for support of a $200,000 appropriation to procure the planes and support facilities, asked, "Why all this fuss about airplanes for the Army? I thought we already had one!"

Knowing I had to fight this kind of apathy, and that the future of flight lay in the heavier-than-air machine, I wanted to do my part to prove what it could do for the Army. I hurried along the modifications to *No. 1*, and at the same time negotiated with the newly formed Army Maneuver Division in 1911 to tie in with their upcoming exercises along the Mexican border. I got little sympathy, and even less cooperation, when I tried to obtain official inclusion in the exercises for some reconnaissance and message-carrying missions. Insult was added to injury when the troops being assembled for the maneuvers moved onto the mounted drill field and left me with only about 10 per cent of the area for my test flights.

I was determined that this fencing-in operation would not stop my flying. When we finally got *No. 1* into the air, the restricted area gave me excellent opportunities to practice short field takeoffs and landings.

After getting over my pique at the lack of elation for my presence at Fort Sam from my ground-bound colleagues, I concluded that their attitude was based on ignorance which was not their fault. Most of the officers and men gathering for the maneuvers had never seen an airplane before I took to the air, and once they bad seen it, they dismissed it as an experimental plaything, flown by "that aeronaut" whose funeral they would soon have to attend. I therefore deemed it appropriate as the commander of the United States air arm to carry out a few indoctrination flights for the benefit of the assembled Army personnel, as well as their horses and mules. I guess you might say that this was the first psychological warfare in which the airplane was involved because I certainly intended to influence the "enemy" by psychological means.

I chose a bright, quiet Texas morning for my first indoctrination flight. I roused my men out of bed, warmed up *No. 1* briefly, and took off toward the row of tents that bordered the north side of my fenced-in flying field. The fact that reveille had not yet sounded did not bother me, but I calculated that it might bother my adversaries.

I zoomed over the tents occupied by sleeping officers of the division headquarters staff at about 10 feet. For the next five minutes I cruised at wide-open throttle over the entire troop area at about 50 feet. I ended my flight with a power dive over the headquarters latrine and landed.

Within minutes of this first indoctrination flight I knew that my efforts were successful. The division commander's aide, Lieut. H. Conger Pratt, arrived on the run with a special message intended just for me. He verified immediately that I had aroused considerable interest among the staff of the division headquarters, especially the commander. Unfortunately, I cannot repeat the words Pratt used because of present censorship laws. Suffice it to say that I had not heard some of the words that he attributed to his superiors before. The words I did recognize had previously been reserved for recalcitrant Army mules.

As a parting shot, Pratt told me that I would hear more of the staff's reactions from Maj. George O. Squier whom I had chased out of the latrine in an indecent condition. This remark did concern me a little, since Squier, as the division signal officer, was my immediate superior. However, Squier had been president of the board on which I had served during the Fort Myer tests on *No. 1* in 1908 and 1909, and he probably had surmised what I was doing. He must have forgotten the incident quickly, because he never mentioned it. His failure to do so was all the encouragement I needed to continue my psychological warfare tests.

This initial test was followed a few days later by a similar test involving a battery of horse-drawn field artillery. Having devoted some time to checking on the drill schedules, I slipped into the air between drill periods and loitered around over the encampment area until I observed a battery moving onto the drill field in a column. When it arrived at the center of the field, I swooped down its flank at about 25-feet altitude. As if by command, the battery immediately dissolved into individual gun units and scattered to the cardinal points of the compass at full gallop. The nearest thing I can think of to describe it is a Roman chariot race where the object would be to see who could get out of the Coliseum first. I considered it an excellent opportunity for the artillerymen to test the reactions of their mounts and themselves under a sudden attack by an enemy aircraft.

Upon reaching the other side of the field, I cut the engine off for a landing. An instant later I noted that I had too much speed and could not make my intended landing spot. Instead, I was headed directly for a line of tents bordering the west end of the field, occupied by staff officers of the Engineer battalion. Fortunately, I had sufficient flying speed and space to make a sharp left turn paralleling the tent line. I was so busy manipulating the controls that I failed to see a horse and buggy standing directly ahead. I still had enough momentum to swerve slightly, but not enough to avert a collision. For a split second the horse and the pilot of *No. 1* glared at each other, and then the horse reared upon his hind legs, upset the buggy, and galloped off to join the artillery horses being brought under control at the other end of the field. The dust the horse stirred up, plus what I made as a result of my strange landing, clouded the air considerably. When I was sure that all motion of the plane had ceased, I unbuckled my seat belt, dusted myself off, and climbed out of the slightly bent air frame.

As my men rushed up to roll *No. 1* toward the hangar, the owner of the buggy, the camp's civilian newspaper distributor, appeared and loudly threatened to sue me for damages to his horse and buggy. His conversation became exceptionally loud. I understood his anxiety over his loss, but I considered the damage to *No. 1* and the near injury to its pilot to be of slightly more importance, particularly since I knew I was going to have to pay for its repairs.

I confess that I may have used some mule language in reply, and I may have done so in a loud voice. In any event, the battle of words ceased when a commanding voice boomed out, "Benny, what's going on over there?"

Turning around angrily, I saw the soldierly figure of an old baseball teammate of Fort Leavenworth school days, Capt. Douglas MacArthur, adjutant of the Engineer battalion. He was standing in the entrance to the tent which would have been torn to ribbons had I not swerved to avoid it. "Doug, I either had to mow your tent down or this man's horse and buggy," I replied.

MacArthur came over to me, looked around the plane, and then said, "Benny, speaking as a disinterested bystander, I'd say you made the right decision."

The newspaperman, seeing he was going to get no sympathy for his case, stomped off toward his horse and buggy in disgust. Doug and I talked for a few minutes reminiscing about Leavenworth days, and parted. As I look back on that incident, I realize that had I been unable to avoid hitting that tent, I might have altered the course of history. I might also have been forced into involuntary retirement many years later, when I, as chief of the Army Air Corps, was fighting my battles on the Potomac. MacArthur alone, as Chief of Staff of the Army, stood between me and my military, political, and industrial enemies in their continuous but unsuccessful attempts to lift my official scalp.

After my loyal crew repaired *No. 1*, I continued my indoctrination course whenever the opportunity arose. While I had been carrying out my own private psychological warfare, a strange development took place that was to mark the first time in history that a private citizen saved an air force from extinction. Robert F. Collier, owner of Colliers Weekly magazine and an enthusiastic aviation promoter, bought a 1910 *Wright-B* and loaned it to the Army pending the availability of funds for the purchase of new planes.*

It arrived at Fort Sam on February 21, 1911, along with Philip O. Parmalee, who was one of Wright's newly trained pilots, hired by Collier to check me out on the new and different control mechanism. This arrival enabled me to put *No. 1* in a "tactical reserve" status, where it remained until I got orders to ship it to Dayton for restoration to its original condition as a Wright Type-A, Model 1909. Eventually, it was placed on the retired list and given to the Smithsonian Institution where it may be seen today.

Collier's gesture as a public-spirited citizen was appreciated, and after a few flights to check me out, Parmalee and I entrained for Fort McIntosh at Laredo, Texas, on February 27, with the Collier *Wright-B* stowed in the baggage car. I had finally obtained permission to accompany troops on maneuvers but only with the understanding I would not "interfere" in any way with the planned exercises.

After a few flights to service test the plane at Laredo, Parmalee and I took off on the first official military reconnaissance flight in history on March 3. We were ordered to reconnoiter along the Rio Grande River from Laredo to Eagle Pass, a distance of 106 miles,

* Collier is remembered in aviation circles today as the donor of the beautiful Collier Trophy awarded annually by the National Aeronautic Association "For the greatest achievement in aviation in America, the value of which has been thoroughly demonstrated by actual use during the preceding year."

and look for "enemy" troops. We completed the flight in two hours and ten minutes at an average altitude of 1,200 feet but never saw the enemy. However, since we carried a total weight aloft of 1,400 pounds, we established an American record for that weight and distance.

The terrain over which we flew was the roughest I had ever seen. My stomach grew large knots several times when the engine backfired and threatened to quit. Once we got on the ground at Eagle Pass, Parmalee and I concluded that we had used up our luck, and decided to take a safer route back. Two days later we took off from Eagle Pass, but decided to follow the Rio Grande where occasional sand bars, or the water itself, would give us a better chance of survival if we should have a forced landing.

I was flying this time with Parmalee acting as observer. About 25 miles out and at about 75 feet of altitude, Parmalee sighted some ducks. In a jovial mood, he threw up his arm and sighted shotgun fashion at the birds. In doing so, his arm hit the engine cutout cord and the engine went dead. I jammed the nose down to maintain flying speed.

"Cutout!" Parmalee yelled, and grabbed for the controls. I grabbed for the cutout lever and pushed it back into firing position. Neither of our reactions could save us. Just as the engine started to fire, the plane smashed into the water and flipped over on its back. The river was only four feet deep where we hit, but it might just as well have been a thousand. Both Parmalee and I were pinned underneath. "What a lousy way to die," I thought. "An airman should die in the air, not under water. Maybe this is what we get for flying on Sunday."

Parmalee and I extricated ourselves after what seemed to both of us an eternity, and splashed to the surface. Parmalee let loose a string of profanity as soon as he could breathe that rivaled any Army mule skinner who ever lived. He didn't stop until I grabbed his sleeve and pointed upward. There, coming from all points of the compass, an enormous fleet of Texas and Mexican buzzards was gathering. Parmalee ceased his tirade abruptly. The full impact of our predicament got to him, and he said, "Foulois, this could be serious. Are you all right?"

I said I was, but admitted we had a problem because it was a long, hot walk back to Eagle Pass. We were sitting on the plane skids when suddenly a voice called out, "What're you boys doin' down there?" It was a Texas cowboy who had seen our plane dive into the river and had ridden over to make sure he had seen what he thought he saw or confirm that he had been out in the hot sun too long.

We asked our welcome friend to ride to Eagle Pass to notify Captain Van Way, the Army commander there, of our predicament and to send some men and an escort wagon to help us get the plane out of the water. He did and they did. The plane was not damaged much in the crash but the men sent to help us caused some damage mounting it on the wagon. We had to ship it and ourselves back to Fort Sam. And thus ended the first military aircraft reconnaissance in history.

Although we never once sighted the "enemy," there was a compensation. During the preliminary test flights at Laredo, James H. Hare, a photographer for Collier's, arrived. Parmalee took Hare up on several flights and the latter took a number of excellent photographs of the rough and broken terrain along the border in the vicinity of Laredo. This photographic reconnaissance marked the first use of a camera from an aircraft and established the beginning of aerial map making.

There was still another benefit that was derived from the Laredo to Eagle Pass mission. I had had my men install a small radio for communication with Signal Corps stations along the border. On the outbound flight I communicated with those stations along the route to report our progress and send my negative reports on "enemy" activity. While this communication from an aircraft in flight to the ground was not the first time it had been done, it was the first practical use of radio and marked the true beginning of the development of air-borne radio communication in our military history."*

Parmalee and I returned to Fort Sam Houston and our men repaired the Collier machine quickly. On March 17, still anxious to prove what the airplane could do, Parmalee and I asked for General Carter's permission to see how fast we could take a message from his headquarters at Fort Sam to Major Squier at Leon Springs Camp, 26 miles away. He readily agreed and wrote out a greeting to Squier. Parmalee and I took off, delivered the message, and returned with a reply in one hour and forty-five minutes. Although certainly no feat by today's standards, this simple demonstration convinced General Carter and his staff, and some doubting Thomases along the Potomac, that maybe these noisy contraptions might prove useful to the ground forces in some way.

Toward the end of the month Parmalee received a letter from the Wrights, recalling him to Dayton. I could not understand why, but the rumor was that the Wrights had found that our crash in the Rio Grande had occurred on a Sunday, thus violating one of their self-imposed rules not to fly on the Sabbath. I don't know whether this was true or not. The more likely reason was that they needed him to demonstrate their new planes around the country.** They wrote that they were sending Frank Coffyn in his place, another of their company-trained pilots; who would be coming along with the Army's newly purchased second Wright plane, a brand-new *Model B*.

On the same day that Parmalee and I had made our flight from Laredo to Eagle Pass the annual War Department appropriation for the fiscal year 1912 was approved by the Congress. This bill, much to the delight of the few of us truly interested, included $125,000 specifically for aeronautics. Twenty-five thousand dollars was made immediately available. To his eternal credit for fighting for it and winning his battle on Capitol Hill, Gen. James Allen immediately placed orders for five new planes. Three of them were to be of the Wright biplane type and two of the Curtiss single-seat type.

* John A. D. McCurdy, a Canadian, sent the first wireless message from an aircraft in flight to the ground in a demonstration at Sheepshead Bay, Long Island, on August 27, 1910. This achievement raised no interest in the Signal Corps because my indoor tests in Chicago the previous January had already proved two-way radio communication was possible.

** Parmalee was killed a few months later during an exhibition flight.

The mention of the Curtiss aircraft is significant at this point because Glenn Curtiss had made valuable contributions to aviation by this time. His engines had been used on dirigibles successfully, and he had become interested in both lighter-than-air and heavier-than-air machines from his association with Tom Baldwin, Alexander Graham Bell, and the Wrights. He designed, built, and flew his own planes, the first really successful one being the *June Bug*, which won for him the Scientific American prize on July 4, 1908. On May 1, 1910, he staged a spectacular long-distance flight from Albany to New York City, a distance of 142½ miles in two hours and fifty minutes flying time. This won for him the New York World prize of $10,000 and established Curtiss as a leading competitor of the Wrights in the manufacture of airplanes.

Curtiss had established the first flying school in the United States at Hammondsport, New York, in 1909 to train pilots in the use of his machines. He opened a second school in San Diego on North Island in the winter of 1910-1911, and invited both the War and Navy Departments to send one or more officers for flying instruction in his machines free of charge. The Navy promptly Sent Lieut. Theodore G. Ellyson, who became the first official Navy pilot. The War Department did not respond so promptly, but finally did authorize Gen. Tasker H. Bliss, commanding the California Department to send "as many officers as can be spared." Thirty officers volunteered and three were chosen: Lieut. Paul W. Beck, Signal Corps, Lieut. G. E. M. Kelly, and Lieut. John C. Walker, both Infantry officers. These three officers, plus Lieutenant Ellyson and three civilians, constituted the first Curtiss class and began training in February 1911. In April Beck, Kelly, and Walker were ordered to join me at Fort Sam Houston and await the arrival of the new planes so they could continue their training. In my opinion, they were still in the "danger period" which I define as that time in a pilot's life when he thinks he knows all there is to know about flying and will take unnecessary risks while trying to prove to himself and his contemporaries that he is God's gift to the aviation world.

While their training had been going on, Curtiss had designed a new plane which he designated "Type IV, Military." It weighed 700 pounds, had a 50-horsepower, eight-cylinder engine, and could carry a passenger sitting on the lower wing behind the pilot. The first of these was approved for purchase after demonstrations in Washington by J. A. D. McCurdy, and shipped to me at Fort Sam Houston. Eugene Ely, one of the Curtiss civilian pilots, arrived to continue the training of the three new officers in this new version that none of them had flown. This machine, the second bought by the Army, naturally received the designation "No. 2." The combat air arm was indeed expanding, and I was no longer the lone officer trying to keep it alive.

The arrival of the three officers from San Diego constituted a requirement for them to be assigned to some unit. Major Squier, the Division Signal Officer, recognizing the need to keep us together, issued an order on April 5, 1911, organizing a "Provisional Aero Company" with the senior man in military rank, Lieutenant Beck, in command.

Up to this time there had been no necessity to have any formal, written rules and regulations to guide air officers and their mechanics. The orders I gave to my ten men were verbal. There had been no need to issue written instructions on the care and maintenance of air machines, since we had only one. And there was no requirement to prepare any written instructions for the training of student pilots. But now things were different. Major Squier directed me to put all instructions down in writing for the benefit of the new officers assigned and those who would eventually follow. As a result, I wrote the "Provisional Aeroplane Regulations for the Signal Corps, United States Army, 1911," the first regulations of any kind ever issued for a military air unit. In the light of present-day flying it is amusing to see those first elementary regulations. Here are some extracts:

> … There will be but one pilot to an aeroplane, and he will be held responsible for its care, maintenance, and good working condition.
> … An aeroplane should, at all times, be handled carefully, and all members of an Aero section should be thoroughly instructed in its construction and repair.

> … A Pilot, operating a passenger-carrying aeroplane, will have exclusive command as regards the operation of the aeroplane. He will determine what altitude will be maintained during flight, and will select the landing places. If the aeroplane is so designed that a Pilot may be relieved from the actual operation of the machine while in flight, the regular Pilot will continue to have exclusive command over the operation of the machine, and if an emergency arises, he will instantly take over the actual control of the machine.
> … Under field service conditions, every aeroplane should have a Pilot and an Assistant Pilot, as members of its crew, to provide against the possibility of one becoming disabled from any cause, and to provide for a relief on long flights.
> … All Pilots and Assistant Pilots should not only be thoroughly trained in the expert operation, care, and maintenance of their machines, but should also be thoroughly trained in military art, and the theory and practice of wireless telegraphy. They should be thoroughly conversant with map reading, map making, and should have good training in the study of terrain as viewed from an aeroplane at different altitudes.

With this advice once published as a guide, I then wrote the first technical orders pertaining to military aircraft inspections:

> … Immediately after the completion of any repairs or alterations to an aeroplane, the machine will be thoroughly inspected by the Pilot and Mechanic, and will be given a careful test, in actual flight, by the Pilot. The inspection will not only include all parts repaired or altered, but will also include the entire machine, in order that nothing will be overlooked. This complete inspection is particularly important after repairs due to hard landings or accidents, as other parts of the machine may be wrenched or strained.
>
> … Immediately preceding every flight, an aeroplane will be carefully inspected by the Pilot and Mechanic. Each inspection to be made independently of the other. Upon completion of both inspections, the Mechanic will report the result of his inspection to the Pilot; the Pilot will then report the result of both inspections to the senior officer present on aeronautical duty.

I followed the above general rules with more exact inspection instructions for the *Wright Type-B* machine regarding the warping controls, propellers, tail assembly, struts, levers, wheels, skids, and the engine. Then "Provisional Drill Regulations" were written which were nothing more than an application of the "by-the-numbers" Army drill ground methods of instruction. For example, I prescribed the "normal unit" for handling the Wright machine as nine men: pilot, assistant pilot, civilian mechanic, and six enlisted men. The following were typical of the instructions which I felt necessary to prescribe to satisfy my ground-bound superiors:

> … The enlisted members of the Section will be numbered from 1 to 6 and should not be changed unless absolutely necessary.
> … Numbers 1 and 2 should be men of experience, preferably the two senior noncommissioned officers. Number 3 should be one of the lightest weight members of the section.
> … For all formations with the aeroplane, the Section will "Fall In" in single rank, 3 paces in front of the center of the aeroplane, with Number 1 on the right.
> … At the last word of the command, "Prepare to move out," Nos. 1 and 2 will grasp their struts, No. 1 with the left hand and No. 2 with the right hand. No. 3 steps up on the front left skid, with his right foot close to the diagonal brace and left foot on cross brace, steadying himself against the diagonal brace. Nos. 4 and 5 will grasp the bottom horizontal tail pieces with their left and right hands respectively, raising the tail until the skids are approximately level. No. 6 will remain in the vicinity of the hangar.

These instructions were followed with marching instructions to follow such commands of the pilot as "move out," "forward march," and even "prepare for flight, march." At this latter command, Nos. 1 and 2 were to move at double time around their respective ends of the machine and halt in the rear of the propellers. No. 3 was to follow No. 2 and halt three paces to the rear of the center of the elevator. Nos. 4 and 5 were to sit down in the rear of the tail and grasp the tailpieces firmly with both hands. The mechanic (a civilian) was to "move promptly by the right flank of the aeroplane, halting in front of the seat braces, facing the machine."

All of these drill-ground instructions looked good on paper to non-flying officers, but were not very practical, and I knew it. Although we practiced these things a few times, it goes without saying that we certainly never tried this by-the-numbers approach to actual flying.

I followed these instructions with words of wisdom about the care and maintenance of the Wright engine and machine which I had literally learned the hard way. Emboldened by the acceptance of these rules by the non-flying people I had to please, I drafted a set of "Flying Safety Rules," most of which are still sound today:

> … Place the machine in such a position that you will always *rise against* the wind. If the wind is blowing under 10 miles per hour, a start may be made with the wind on the flank. If a start is made with the wind on the flank, great care must be taken in getting off, that the machine does not tilt to one side or the other, as a wing may hit the ground. Also a wind blowing from the flank tends to turn the machine out of its course, and make it run before the wind. In the present types of aeroplanes this is obviated by turning the vertical rudder toward the flank from which the wind is blowing.
>
> … Never start with the wind blowing from behind the machine, if it can be avoided, as it requires a much longer run, before the machine will rise.
>
> … Never land with the wind blowing from behind the machine, as one is liable to underestimate the additional speed, and be carried into obstructions on the ground, which generally results in a wreck.
>
> … Always run engine a few minutes before a flight, to ensure its good working condition and to warm it up.
>
> … Never fly over terrain which has no good landing places, unless absolutely compelled to do so. Whenever this condition becomes necessary, fly at an altitude that will permit, if possible, of an easy glide, with engine dead, to a safe landing place.
>
> … Never attempt exceptional or dangerous evolutions with the machine while in flight.

> … Never start off a flight until absolutely certain that every part of machine has been thoroughly examined, and every part functions properly.
> … Never make a flight in high winds or in very bad weather unless he is absolutely certain of his ability to control the machine, and has had practical experience in operating an aeroplane under such weather conditions.
> … Never fly over towns, villages, or cities unless at an altitude sufficiently high to enable the machine to glide clear of the houses. It is much safer to make a detour and avoid the above, and in case the motor stops, the Pilot will have a much better chance of making a safe landing.

I made many more observations intended to help my fellow aviators, some of which were more humorous than astute. For example, I wrote that "misjudging the distance to the ground in making a landing is usually followed by a wreck and possible injury to the passengers." I also said that pilots should "learn to know when the engine sounds right," and "that great care should be exercised in flying among high mountains or over mountain ranges," and that when "a thunderstorm makes its appearance in the vicinity, the safest policy is to make a landing and secure the machine against damage from the approaching rain and wind."

I wrote quite a discourse on the subject of meteorology, taking much of my material from published literature of the day and translating it into what a pilot sees and then giving my best advice based on my personal experience. I concluded that "it is well to impress upon all officers and men on duty in connection with the aeronautical work of the United States Army that 'eternal vigilance is the price of safety' and that it is always best to 'make haste slowly.'"

I make no claims that I was a man of great vision laying the groundwork for great aerial armies that would soon fill the skies. I wrote what I thought young pilots must know in order to survive. The advice was based on common sense in dealing with machines that can kill or maim. Unlike ground machines, first mistakes were too often last mistakes in air machines. I had to do what I could to minimize ignorance.

When the new *Wright-B* and the Curtiss arrived, I saw the differences in the two machines for the first time. Glenn Curtiss had installed a wheel on a post which, when pushed fore and aft, operated the elevators and when rotated left or right, controlled the rudder. Curtiss had installed ailerons as a substitute for the wing warping of the Wrights. These were actuated by shoulder controls so that if a pilot wanted to bank to right or left, he would lean in these directions. Pilots qualified on the Wright planes could not fly the Curtiss planes and vice versa because of the difference in the controls. This difference remained for several years until aircraft manufacturers finally agreed on a standard control system for the benefit of all pilots.

The difference between the two machines sparked a rivalry between their pilots. Coffyn and I were determined to show that the *Wright-B* was better than the Curtiss and laid out a series of demonstration flights to prove the military potentialities of the *Wright-B* for the benefit of our own maneuver division high command as well as a number of foreign visitors, including military attaches from Germany and Japan. One of these demonstrations was an hour-and-a-half flight that was announced to take place "regardless of weather conditions." Coffyn and I made the flight as scheduled in a steady rain which caused us to take off from a soft field covered with about six inches of sticky mud. We flew around for an hour and a half and proved to ourselves and our observers that an airplane could fly in the rain, a fact often disputed in those days.

The three Curtiss-trained officers were eager to prove what *No. 2* plane could do, but it frightened me to watch them operate. They were not fully checked out on the Curtiss. It was a one-man machine even though another person could ride the rear seat. There were no dual controls so these three pilots generally flew solo and had to teach themselves to fly. They did many things that I had learned not to do, and practiced their mistakes time and time again.

Lieut. John Walker scared me first shortly after he got there. While attempting to make a steep turn to the right at about one hundred feet altitude, he stalled and dropped to a few feet from the ground where he recovered briefly and then banged into the ground for a very hard landing. This landing didn't only scare me, it scared Walker, too. He never flew again and subsequently asked to be relieved from flying duties.

The next day Lieutenant Beck took off in *No. 2* and, instead of maintaining a safe altitude over or around the field, hedgehopped over the mesquite in the local area. When he decided to land, he dragged the plane in just over the treetops at low throttle. The engine quit before he reached the field boundary, and he ended up in a heap about a hundred yards short of the field. The plane was badly damaged but Beck escaped unhurt. He was knocked out briefly and when he came to, he found himself walking around through the brush with the control wheel in his hands.

The repair work, under Beck's direction, began immediately and consisted mainly of replacing the front steering wheel fork assembly and the main gear. I watched this repair work with interest because it gave me a good chance to see just how the Curtiss machines were put together. But what I saw I didn't like. One of the two wooden forks was knotty, cross-grained, and of inferior quality as compared to the knotless, straight-grained quality of wood used in the Wright planes. I also saw where an unnecessary bolt hole had been drilled in the same fork, probably because the mechanics made a mistake in measurement. I called this possible hazard to Beck's attention but he shrugged it off.

On May 10 Lieut. G. E. M. Kelly was scheduled to fly this plane for his first qualification flight. The least experienced of the three pilots, he climbed in *No. 2* and took off. No test flight had been made to check the newly installed wheel assembly.

Kelly came in hot and high on his first landing attempt. To get the plane down, he put it at a steep gliding angle and smacked into the ground on the nose wheel, then the main wheels, and bounced back up into the air again. He still had sufficient control, however, and pushed the throttle wide open to go around for another try.

The second approach was no better than the first. My heart stood still in my chest as I watched him dive steeply toward the ground and hit nose wheel first again. There was a sickening, wrenching sound on impact and I thought the plane would disintegrate right there. But it didn't. Instead, it bounced skyward about 30 feet and headed directly for the line of tents on the far side of the field.

Kelly must have sensed he wasn't going to be able to go around this time. With the tents only about 75 yards in front of him, he made an abrupt left turn. The steep angle of his wings told me he would never make it. He didn't have enough speed to make a steep bank, and gravity took over. In the next two seconds *No. 2* was a mass of splinters, cloth, and metal. When it hit, Kelly was catapulted out of the seat and landed on his head about 20 feet away. He died a few hours later of a fractured skull.

Kelly died a needless death, in my opinion, and when Major Squier asked me to serve on a board to investigate the accident, I told him I would serve if ordered, but I would be a prejudiced member. I then told him what I had observed while *No. 2* was being repaired under Beck's supervision and that I had given Beck my opinion as to the possible hazardous condition of the plane before Kelly's flight.

My bitterness came through in my discussion with Squier and I told him that if I were called as an eyewitness before the board, I would name Beck responsible for his failure to carry out his technical and command responsibilities by not testing the plane before Kelly's flight, and by not having the professional competence to realize a potentially unsafe airplane when he saw one. In my anger, I told Squier that in view of my three years' practical air experience as compared to Beck's five months, I had lost all confidence in his qualifications to be the commanding officer of the Provisional Aero Company and that one of us should be relieved.

Squier did not put me on the board and I was not called as a witness. Thus, my prejudiced viewpoint was not introduced into the investigation. Without my views, the final report in the official files reads as follows:

> … He [Lieutenant Kelly] had made a not abnormally hard landing. Upon landing at least one and possibly both sides of seat fork were broken at a point between pilot seat and foot rest. At the same time

> it appears that one diagonal bamboo brace from front wheel to front elevator was broken, and its mate was bent.
> … It is the unanimous opinion of the Board that the front wheel must have struck an abrupt depression in the ground or some obstacle causing the strain which resulted in the break.
> … From all of the preceding facts, the Board is of the unanimous opinion that the accident was due to the efforts of Lieutenant Kelly to avoid endangering the occupants of the Eleventh Infantry Camp in which endeavor it became necessary for him to make a sharp left turn, which, in the crippled condition of the machine, put more strain on the controls than should have been required in a straightaway landing. Such straightaway landing was impracticable owing to the proximity of the tents.

The board then listed the probable cause of the accident as "error in judgment on part of pilot in selecting an unsuitable landing place."

My conclusions are, entirely different. Kelly died because he was poorly trained, inexperienced, and was flying an unsafe airplane. Unfortunately, these were things he was personally unaware of, and so was his commanding officer and the other members of the investigating board. It was this same kind of basic ignorance about the essentials of aircraft maintenance and flying that would kill many more pilots in the years to come.

The death of Kelly had a profound effect on General Carter. After Kelly's funeral, he issued an order prohibiting further flying from the Fort Sam drill field. The Provisional Aero Company, barely a month old, was disbanded and all personnel and equipment ordered to move to the newly established aviation school at College Park, Maryland, under the command of Lieutenant Beck. That is, all personnel were to move except one officer—me. Because I had openly clashed with Beck, I was assigned to a desk job in Washington in the Division of Militia Affairs of the War Department. This was the third wound stripe received during my military career for sticking my chin out and saying things that were not popular. It was a bitter experience, but I had only one regret. I had learned to love the smell of airplanes, the blast of wind in my face, and the exhilaration that only flying can give a man. As I packed my bags for the trip to Washington, I wondered if I could take being chained to a desk, especially when it seemed to be a punishment. For the first time in my military career I seriously considered resigning my commission and going back into civilian life.

Chapter Seven

The Baptism of American Combat Aviation

The initial funds voted by the Congress for aeronautical activities for fiscal year 1912 stimulated the establishment of a flying school at College Park, Maryland, the purchase of five planes, and the acquisition of students to fly them. Lieuts. Roy C. Kirtland, Henry H. "Hap" Arnold, and Thomas DeW. Milling reported for duty, the latter two after receiving dual instruction from the Wright school in Dayton. Capt. Charles DeF. Chandler was named commanding officer. Both he and Kirtland received their instruction from Arnold and Milling in the following months. Beck, now a captain, joined the group in June, 1911.

Once settled down, the five newest pilots in the Army inventory began to make history. Flying was still a novelty and any flight of any distance aroused much publicity, especially if there was any advance notice.

In August 1911, the first "long cross-country flight" was announced when the Army was invited to send two airplanes from College Park to visit the annual encampment of the District of Columbia National guard near Frederick, Maryland. The distance was 42 miles. Arnold and Chandler got there, but Milling and Kirtland had a forced landing near Kensington because of engine failure. On the way back Arnold and Chandler got lost and landed to ask directions from a farmer. In trying to take off, they crashed, and had to come the rest of the way by train.

Hap Arnold set an altitude record for military planes on July 7 when he reached 3,260 feet on a local flight from College Park. He beat his own record on the eighteenth when he ascended to 4,167 feet.

While these records were important as far as we in the military were concerned, the civilian fliers were also setting and breaking records and making precedents. In August Harry Atwood flew from St. Louis to New York—1,266 miles—in the longest cross-country flight up to that time. Lincoln Beachey set a new altitude record of 10,837 feet at the Chicago Air Meet on August 20, and on the seventeenth Calbraith P. Rodgers began the first transcontinental flight from New York. It took Cal forty-nine days, thirty stops, and twenty-five crackups to make it, but he did.

Not to be outdone by either the Army or civilians, the United States Navy purchased its first plane in May, 1911, and attempted the longest overwater flight in history in October. Lieuts. J. H. Towers and T. G. Ellyson flew from Annapolis, Maryland, to within two miles of Fort Monroe, Virginia, in two hours, twenty-seven minutes.

The fever of setting new records and achieving aviation "firsts" was catching everywhere: The five-pilot air force at College Park set more of their own. In September Tom Milling became the first Army pilot to fly at night when he landed his plane by the light of gasoline flares at an air meet in Boston after participating in an air race. He also piloted a *Wright-B* on a three-man endurance flight on September 26 and won the Rodman Wanamaker Trophy.

In the strictly military sphere, Riley E. Scott, a former Coast Artillery officer, invented a bomb sight in 1911 and brought it to College Park for testing. Weighing about 64 pounds, it was a small telescope affair which served to determine the actual speed over the ground by solving the right-angled triangle with two known sides—altitude above the ground taken from an altimeter reading and a stop-watch reading taken when passing over a previously observed ground position. Logarithmic tables were prepared for 400-feet altitude so that no in-flight calculations were needed.

During the tests, Scott lay prone on the lower wing with the sight fixed on the leading edge. Eighteen-pound bombs were dropped within ten feet of a 4 feet by 5 feet target.

The rapid breakthroughs being made in the science of flying did not go unnoticed in the Signal Corps. Toward the end of 1911, it was recognized that the military planes of the future would have to have longer range and greater weight-carrying capability if they were to be of any value with ground troops. It was widely accepted then that airplanes had only two military purposes: aerial reconnaissance and rapid transportation of senior ground officers from point to point. Even though bombs had been dropped and guns fired from planes, no one took these efforts as giving any promise that the airplane could ever become an offensive weapon.

To get better planes, the Signal Corps consulted with the several plane manufacturers to determine what specifications might reasonably be laid on the industry. It was agreed that speed and a high weight-carrying capacity could not be combined in the same plane with the low-horsepower engines then in production, so two different types of planes would be required. One of these, designated the "Scout" version, was needed for reconnaissance and would have to be designed to carry two pilots and a heavy load of gas, radio, and photographic equipment. The second version, called the "Speed Scout," would be a single-seat aircraft designed for strategic reconnaissance and rapid message delivery.

The "Scout" specifications called for dual controls, a four-hour· endurance, and a top speed of 45 miles per hour. The "Speed Scout" had to carry enough gas to fly a radius of 100 miles at a speed of 65 miles per hour. In January, 1912, orders were placed for five new planes meeting these specifications. Four of them were to be "Scouts"—two Wrights, one Curtiss, and one Burgess-Wright with a Renault engine. The fifth was to be a Wright "Speed Scout."

In order to continue flying during the winter of 1911-1912, the decision was made to transfer the flying-school operations from College Park to Augusta, Georgia. The officers and the four planes were taken there by train and flying began on December 7. The Weather Bureau had provided data which showed that this area had a low average wind velocity, no snowfall of any significance, mild temperatures, and little rainfall. But the weatherman didn't know what the winter that year was going to be like. Two heavy snows in January and February, excessive rains and floods, changed the averages, and the four-plane air force was under water most of the time. In the one hundred and twenty-four days that the group was there, only eighty-one hours of flying time was logged. Flying was resumed at College Park in April, 1912.

While the art of flying was advancing without me as an official contributor, I settled down to my job in the Division of Militia Affairs. I was an eternal optimist and had decided that if I were right in my views about aircraft maintenance and flying, time would heal the wound I had suffered; my superiors would correct the injustice they had done in time, and let me get back to flying.

My job was to maintain staff supervision of the organization, training, and operation of Signal Corps and Engineer units of the National Guard of the forty-eight states.

In October 1911, I passed out at the breakfast table and was carted off to Walter Reed Hospital. The diagnosis: acute appendicitis. During the operation, the surgeon, Dr. H. Wellington Jones, had a tough time keeping me on this side of the "Great Divide." A nurse later told me that my heart had stopped beating twice during the operation and that the doctors had brought me back both times by artificial respiration.

"Lieutenant Foulois," the nurse told me afterward, "there's only one reason you're alive. When your name was called out at the pearly gates, you were refused admittance, and when your name was called out at the devil's door, you were turned down again. They wouldn't have you in either place!"

I guess I did almost file my last flight plan at that time because when I was able to navigate in a wheel chair, the medic called me to the window one day and pointed outside. "See that building there, Lieutenant? That's the hospital morgue. We had a locker all reserved for you in there for the first ten days!"

The news of my expected demise got around the War Department quickly via the Washington newspapers which ran headlines about "Nation's Foremost Airman Not Expected to Live" and ran an advanced obituary accounting of my contributions to aviation. It was nice to know that someone was giving me credit for the things I had done, but it was uncomfortable reading my own obituary.

Two of my Signal Corps colleagues evidently were already referring to me in the past tense because they both submitted formal applications for assignment to my job. These came to the attention of Gen. Leonard Wood, Army Chief of Staff, who immediately summoned my boss, the chief of the Division of Militia Affairs. The air is alleged to have turned dark purple as the general gave my boss his own personal brand of mule language. "You could have at least held onto these applications until Foulois died," he shouted. "Now take these back and tear them up and don't *ever* try to fill a man's boots until he's pronounced dead!"

I learned of this episode much later, and when I did it upset some of my ideals about the brotherhood of officers assigned to Washington. From that time on I vowed that I would do my best to stay in field duties with clean-fighting combat soldiers rather than serve in the back-stabbing atmosphere of the Potomac River battlefront. I shall never forgive those two officers for their thoughtlessness in trying to get my job before I vacated it.

One of the many things that I had to do something about was a proposal to organize air units within the National Guard structure. The states of New York, New Jersey, Pennsylvania, Connecticut, Illinois, and Missouri had all indicated interest in starting air training of some kind, but all proposals were turned down because it was War Department policy to give priority to the development of Signal Corps ground units. But it was the "show me" state that broke the barrier and it was my poor poker playing that enabled Missouri to claim the first air units for their National Guard.

I had determined that once I had whittled down the paper pile I would go on the road to crusade for the establishment of National Guard air units, because I recognized the political pressure that guardsmen had on the Congress and the War Department. I saw that Missouri had already established an unofficial air section and had taught its few members about the rudiments of flight and ballooning, so I went to Jefferson City, Missouri, on an inspection trip. I reported to the Adjutant General, and made an inspection of the Signal Corps unit, accompanied by his staff. We then went to dinner where I was asked casually if I would like to check on the brand of poker played by the staff. Being an indifferent player, I decided that the game would give me a chance to keep on working while they played.

The game began, and my hosts began to prove their prowess at my expense. However, between hands I talked flying and kept up a running monologue about the thrill of flying, the future of the airplane, and how foreign governments were leaping ahead of us technologically. When the poker party broke up, I was $75 behind the game but I had a verbal IOU from the Adjutant General that he would do everything in his power to create an official air unit, which he did soon after. This was the beginning of what is now the Air National Guard.

This bit of pioneering in the fine art of political persuasion was my last for a while. I stayed at the job in the Militia Bureau but I frequently visited the Signal Corps machine shop and helped design a radio generator that could be operated by attaching it to an airplane engine. One of these units was sent to Augusta in the winter of 1911-1912 and used at College Park the following summer. The set did not prove satisfactory but I kept on trying in my spare time.

During the months of June and July, 1912, I made arrangements to help two civilian radio experts carry out more tests. I arranged it so that I flew the plane, and by the end of July we were able to transmit wireless telegraph messages from altitudes of 2,500 feet and over a distance of 18 miles.

In carrying out these tests I used a Burgess-Wright two-seater because I wanted to use one of the civilian engineers as the radio operator while I flew the plane. However, after one try I found that I couldn't get the plane into the air with the combined weight of the pilot, the engineer, and the wireless equipment. I continued the experiments alone and sent messages myself by using my thumb to operate the transmitter key located on one of the control levers.

In between my desk work and the wireless experiments I designed an instrument for plotting and map making in-flight, improved the safety belt, and qualified for the new Military Aviator pilot rating.*

By the fall of 1912 a number of new pilots had been trained and all three airplane manufacturers—Wright, Burgess, and Curtiss—were turning out hydroplanes which were land planes equipped with pontoons. Burgess and Curtiss followed the European plan of equipping their planes with an enclosure around the pilots who were seated in tandem. Burgess was now installing a tractor propeller in front of the plane instead of a pusher in the rear.

In the latter part of July 1912, the College Park personnel were ordered to take part in Army field exercises in Connecticut. It was the task of the four-plane air force to test the efficiency of the airplane to work in conjunction with ground troops. Although not on aviation duty, I managed to talk my boss into letting me go along with the College Park personnel. Flying operations began on August 8.

Although four planes were to take part, only three got to Connecticut. Lieuts. Hap Arnold and Roy C. Kirtland cracked up the Army's first tractor hydroplane in Massachusetts Bay as they were taking off from the Burgess factory to join us. They were unhurt, but this left only two Curtiss planes and my Burgess-Wright (the only one equipped with radio) to carry on.

There were seven pilots assigned to our group. We all made numerous reconnaissance flights and during good weather reported the positions of "enemy" ground troops. On August

* This test required the candidate to ascend to 2,500 feet, fly in a 15-mile–per-hour wind, carry a passenger to 500 feet, make a dead-stick landing within 150 feet of a designated point, and make a military reconnaissance cross-country flight of at least 20 miles at an average altitude of 2,500 feet.

14, while flying one of my assigned missions, the engine gasoline feed pipe broke and I found myself looking for a patch of terrain to bring that flight to a successful conclusion. The only field I could see was a small plowed area about 75 yards long and 25 yards wide, surrounded by trees and a stone fence. I hit the ground smoothly enough but couldn't avoid hitting the stone fence. It proved to be an excellent arresting device. The front skid braces slammed into the rocks. If I hadn't had my safety belt on, I would have been catapulted into the next county. But the airplane took the beating and I suffered nothing more than a few missed heartbeats.

I telephoned the results of my reconnaissance to headquarters and asked for a crew to disassemble the Burgess and transport it back to the maneuver camp. While helping the crew lift a large piece of it, I felt my stomach give way in an old-fashioned rupture. I returned to camp, got fitted with a field truss, and was back in the air the next day with the repaired Burgess-Wright.

On August 17 I spent three hours on reconnaissance as the exercises were drawing to a close. During the fast-moving ground operations I had to land frequently to report enemy positions to the commander of the Blue Forces. In late afternoon I reported that the right flank of our Blue troops was being turned by the enemy Red Force. The Blue commander immediately withdrew his troops, but before I could take off, my plane and I were captured by the 2nd Connecticut National Guard. The commanding officer, Col. John Q. Tilson, then a member of the Congress, was a personal friend and paroled me on the field. Once his prize captive was accounted for by the maneuver umpire, I flew back to camp, turned in my flying equipment, and returned to Washington to have my hernia repaired.

Today those Connecticut maneuvers must seem insignificant. During the maneuvers our planes had operated first with the Red Forces and then with the Blue. In each case we were able to give the force to which we were attached a decided reconnaissance advantage over the other. We proved that airplanes could replace the cavalry and could prevent surprise mass attacks by providing information on enemy troop buildups and movements much faster than ever before.

The pilots all received commendations for which we were grateful. When asked for our recommendations as a result of the maneuvers, we agreed on two basic points: (1) trained observers were needed for reconnaissance to relieve the pilot of the dual responsibility of flying the airplane and accurately assessing enemy deployments; and (2) future planes should be designed with a fuselage or cockpit to protect the crew from the wind.

One of the enlisted men who had helped me on the wireless experiments at College Park and on the Connecticut maneuvers was Corp. Frank S. Scott. He visited me at Walter Reed Hospital where I was recuperating from my hernia repair. As he started to leave the room, he stopped, thought a moment, then pulled a horseshoe out of his pocket.

"Here, Lieutenant," Scott said, "is a present for you. It has been my good-luck charm, and I'd like you to have it."

Scott's gesture meant more than words. He was giving me, a former enlisted man, a gift which came from the heart. Tears welled up in my eyes as he left. A few weeks later Lieut. L. C. Rockwell took Scott along as a passenger on a local flight from College Park. After flying for a few minutes, Rockwell was observed making a power-off glide. About 25 feet from the ground he gave the engine full throttle and the plane stalled and crashed. Scott was killed instantly, and three hours later Rockwell died at Walter Reed Hospital.

Rockwell Field, now part of the Navy's San Diego Air Station, was named for the young pilot. Scott Air Force Base, Illinois, is named as a permanent tribute to the first enlisted man to die in furtherance of the development of military aviation. I still have Scott's horseshoe today. It must have indeed brought me luck for I was never hospitalized again for the rest of my military service, and it has an honored place among the Moro knives, swords, and guns on my den walls. It is a symbol that represents the camaraderie between all enlisted men and their officers who work and fight and share the risks of military flying together.

While I had been in the hospital, Capt. Charles DeF. Chandler, the commanding officer at College Park, had recommended to the Chief Signal Officer that I be reassigned to full-time aviation duty. I followed this with a request on my own behalf. Both the recommendation and the request were turned down. But I won half the battle. I received orders to proceed to Fort Leavenworth, Kansas, for duty with the 7th Infantry and to establish an "aviation center," but this had to be done in addition to my other duties. Translated, this meant that I had to do all of my work with the 7th Infantry during the day. What time was left over could then be filled with aviation duties. This kind of thinking on the part of the Signal Corps high command just proved that the Army really hadn't decided that the airplane was going to stay.

I arrived at Fort Leavenworth in November 1912, and immediately began looking for a suitable piece of ground for a flying field. I found what I wanted adjacent to the military cemetery, an appropriate location for flying fields in those days. Once this was decided, I felt honor bound to report to my superiors on the Potomac that all I needed to start an aeronautical center was an airplane, a hangar to put it in, and men to work on it.

Much to my surprise, the post quartermaster provided the hangar, enlisted men were ordered to join me, and I received notice that a *Wright-C*, Signal Corps *Aeroplane No. 10*, was arriving from Fort Riley, Kansas, by train. It did, and what an airplane! It had nearly ended the career of Lieut. H. H. "Hap" Arnold, and was shipped to me in pieces.

The *Wright-C* already had a reputation for its poor flying characteristics, so while my men were making replacement parts I corresponded with Orville Wright to get some advice on how to fly it.*

Orville responded promptly as he always did. However, before I could get *No. 10* back in the air, I received a telegram in February 1913, from the Chief Signal Officer, requesting my recommendations concerning proposed congressional legislation for the establishment of an "aviation corps" as part of the line of the Army. I answered as best I could, not knowing what the proposed bill said. On the same day I wrote to a civilian friend in Washington and asked for a copy of the bill (H.R. 28728). When I saw what it contained, I immediately fired off a long letter recommending disapproval to the Chief Signal Officer.

The Proposal of H.R. 28728 was indeed revolutionary, and it marked the first move to separate aviation from the Signal Corps. However, this kind of legislation was premature, not well thought out, and contained many objectionable features. Since I had been asked for my honest opinion, I gave it. Here is part of my reply:

> The individual and collective experience of all of our military aviators has not, as yet, brought forth sufficient data to warrant the establishment of an Aviation Corps as an independent organization, notwithstanding the fact that foreign armies have done so. Our service and our organizations in the United States Army are comparable only in a very slight degree to foreign standards, and we should not base our views, conclusively, on what is being done by French, German, and English armies.
>
> Legislation is absolutely necessary for the future development of aviation but hasty and ill-considered legislation, such as is shown in this bill, will never result in anything but future confusion, discord, and discontent.
>
> In reading over the bill, it is plainly evident that the views of all the men most vitally interested (the aviators) have not been taken into consideration, nor has the present policy of army re-organization been thoroughly considered.

The one provision of the bill that proved it had not been

* Wilbur Wright had died of typhoid fever on May 30, 1912—a sad day for military aviation.

carefully considered allowed for only two captains to be assigned. There were two captains on aviation duty and one scheduled to return to troop duty, plus six senior first lieutenants, all of whom would be promoted soon. Thus, the limitation on rank would soon throw out the core of experience which had been won at the expense of several lives and planes.

Another provision of the bill was that "no officer shall be detailed as commandant of the Aviation Corps unless he shall have displayed especial skill and ability as a military aviator." My answer to that was that "there is not an individual aviator in the United States Army who has, as yet, displayed especial skill and ability as a military aviator; that is, judging from military standards and not newspaper standards." I included myself in that statement because I figured that there wasn't a single pilot in uniform that was truly master of the flying machine when we could not fly more often than we did.

"The head of the aviation service, for some time to come," I wrote, "should be an experienced executive and administrative officer with a broad general knowledge of both aviation and aeronautics and capable of grasping and coordinating the views and ideas of the men actually engaged in flying." While I fervently hoped that someday in the future there could be a co-equal aviation corps, I did not see how it could be done when we had only a handful of inexperienced pilots, no logistical organization, and little, if any, concept of how airplanes should be used in support of ground troops. We needed time to prove out the airplane and ourselves.

I disliked playing the game of politics, but I decided that since my Washington superiors had dealt me a hand by officially asking me for an opinion on pending legislation, I would play to the best of my ability. I felt it appropriate, therefore, to build a little political backfire of my own. I wanted to talk to Col. John Q. Tilson, the same Tilson who had captured me during the Connecticut maneuvers and who represented my own congressional district. He was an influential member of the House Military Affairs Committee, and I felt he should have the benefit of my reasoning on the subject. Being of the old school, however, I didn't want to go to my congressman via the back door. I took a fifteen-day leave, went to Washington and asked permission of the Chief Signal Officer to see Colonel Tilson. To my surprise, permission was granted eagerly because no one in the War Department wanted the bill to pass either—but for different reasons. Tilson was politely sympathetic and noted wryly that my views made more horse sense than the official comments of the War Department. In spite of Tilson's dissent, the bill was passed by the House but subsequently killed in the Senate.

The visit to see Congressman Tilson had a bonus effect, however. He told me that the Army appropriation bill for fiscal year 1914 was being held up in the Military Affairs Committee because of differences of opinion on the $125,000 item for general aviation, plus a new item called "flying pay" for military aviators. The idea of giving extra pay to military flyers as a compensation for their hazardous duty made the former ground soldiers in the Congress furious. The proposal called for an additional 35 per cent to be added to the base pay of flying officers who were on full-time aviation duty.

The concept of extra hazard pay had the backing of the War Department because of the numerous crashes that always received wide publicity and the many deaths that had taken place in civilian and military aviation up to that time. Six Army men had died by the end of 1912 and the aviation profession already had a bad name with the insurance industry. Anyone who flew found it either impossible to get himself insured or found the rates so high that he couldn't afford it on his Army pay. Flying had become a serious morale problem and extra pay was considered a necessary incentive.

The chief opposition to the flying-pay idea came from Congressman James Hay of Virginia, according to Colonel Tilson. "If you can persuade the old duffer to change his mind, you'll get your flying pay," Tilson told me. With the consent of my Signal Corps superiors, I decided to try.

I had no idea how an Army officer, even with the support

of his superiors, should go about contacting a congressman for the express purpose of changing his mind about pending legislation. But I had the confidence of inexperience. I also had a good memory for names. Out of my mental files, I recalled the names of two maiden ladies whom I had met on board ship while returning from France in October, 1909. During the course of our daily shipboard association, I learned that both of them were government civilian employees and that one of them, whom I shall call Julie, had a very broad acquaintance with Washington politicians. I recalled, to my great delight, that she mentioned Congressman Hay's name many times. It seemed that she was famous for her friendly poker parties and that Mr. Hay often attended to relieve himself of the tensions of his day on Capitol Hill.

Checking through the Washington telephone book (not a difficult task in 1913), I found that both of my shipboard friends were still in Washington. Whereupon, I contacted Julie, made a date, and visited her apartment. I enlisted her help in arranging a meeting with Mr. Hay but without his knowing why.

"That will be easy," Julie said. "All I have to do is tell Congressman Hay that I've got the world's worst poker player visiting me on Saturday night. He'll come even if he has to cancel a dinner at the White House."

Julie was right. The congressman accepted the invitation and we laid our plans. I was to pick up Mary, my other shipboard acquaintance, and bring her to Julie's apartment at eight o'clock. Julie would have the congressman arrive at the same time and in the preliminary chitchat over a drink would bring up the subject of aviation. I was to answer her questions and carefully inject the terrors and hazards of flying into the conversation. We started our friendly game of poker and between periodic breaks between hands and visits to Julie's bar, I found ample opportunity to tell my story of military aviation, under the able and expert questioning of our hostess—interrupted by occasional searching queries from Mr. Hay.

Two hours later, during which I donated $50 of my pay to the congressman's bankroll, I decided that it was time to go. Mary and I departed leaving Julie to carry on the flying-pay argument with the good representative from Virginia.

The following Monday I telephoned Congressman Tilson to tell him of my meeting with Mr. Hay. With a chuckle, Tilson told me that he had just returned from a meeting of the Military Affairs Committee where Mr. Hay had recounted my arguments in favor of flying pay, which he said he had arrived at "through diligent research this past weekend." This "diligent research" resulted in his agreement to support the provision of the appropriation which included the item for flying pay.

Historians record that the War Department appropriations bill with the initial flying-pay item included was enacted into law on March 2, 1913. They do not record that it cost an Army first lieutenant half a month's pay and a gallon of a certain maiden lady's best booze to bring this aviation "first" into being. Neither does it record that the lieutenant who contributed to a congressman's research on the subject was himself ineligible for flying pay because the bill provided only for full-time aviators to be so compensated.

The beginning of 1913 had seen more advancements made in United States military aviation. The Signal Corps now had purchased a total of eleven planes. Of these, eight were at College Park, one had been totally destroyed, one was in the Philippines, and one (old *No. 1*) was hanging from the ceiling of the Smithsonian Institution. In November artillery fire had been directed from an airplane for the first time by Lieut. Hap Arnold. Arnold had also set an Army altitude record of 4,674 feet and had been the first recipient of the Mackay Trophy for winning a contest at College Park. The first Army air school was opened in the Philippines with Lieut. Frank Lahm commanding his one-plane training squadron. Capt. Paul Beck had set an unofficial military endurance record of four hours, twenty-three minutes. Small accomplishments as they seem now, each of them provided steppingstones to the future.

The turn of the year had also produced political unrest in Mexico which had led to serious border disturbances and threatened the peace between that country and the United States. On February 22 General Victoriano Huerta had seized the reins of government after the assassination of President Francisco Madera. Former supporters of Madera threatened to bring civil war to the politically unstable country; and to add to Huerta's difficulties, President Woodrow Wilson, upon assumption of office, adopted a policy of "watchful waiting." Wilson withheld recognition of Huerta because of the manner in which he had come to power and permitted the sale of arms to his chief opponents, Gen. Venustiano Garranza and Francisco Villa. Resentful of this treatment, some of Huerta's followers arrested United States sailors while in the port of Tampico. They were soon released with an expression of regret from Huerta, but Adm. Henry T. Mayo, commanding the United States fleet, was not satisfied and demanded an additional apology in the form of a salute to the United States flag. Huerta refused unless the Mexican flag was saluted in return.

Admiral Mayo, backed by President Wilson, refused Huerta's counterdemand, and the policy of "watchful waiting" was temporarily abandoned. The Mexican port of Vera Cruz was occupied by United States forces after a naval bombardment that cost many lives. War with Mexico seemed inevitable but was avoided when Huerta resigned and Carranza became president. "Pancho" Villa, Carranza's former ally, started another revolution in an attempt to win control of most of northern Mexico. Plans were prepared to shift Army units into position along the Mexican border in case Villa decided to cross it.

While I was at Leavenworth, the school at College Park had closed and moved to Augusta for the second time. Captain Chandler and Lieuts. Harry Graham and William C. Sherman—all Wright pilots—took their five planes and equipment down the coast by rail. Lieuts. Roy Kirtland and Tom Milling followed later, but Hap Arnold remained in Washington as head of the Aeronautical Division. Arnold was still severely shaken by the spin and crash he had had at Fort Riley and it was many months before he actually flew again.

Operations began at Augusta in January 1913, and new planes were tested and added to the inventory—and there were more accidents. Up to the last week in February no landings had been made away from Barnes Farm. Chandler wanted to encourage cross-country training and made the first flight away from the field when he flew to Waynesboro, Georgia, 28 miles south of Augusta, on February 25. He landed in an open field at the home of a friend for an overnight stay. That evening, he received a telephone call ordering him to transfer all personnel and equipment to Texas City, Texas, as quickly as possible because of the seizure of power by General Huerta. Three days later, a special train of eleven cars carrying 5 officers, 21 enlisted men, and the planes with the associated equipment left Augusta.

Flying at Texas City began about March 5. The assembled group, now including two pilots who had been training at Palm Beach and an administrative officer from Washington, formed what they called the 1st Aero Squadron. This designation was not official but was based on a memo from the Chief Signal Officer to the Army Chief of Staff which suggested that "as the present concentration of aeronautic material at Galveston presents the first opportunity to officially recognize an organization," the name "1st Aero Squadron" should be the official designation. However, no general orders were issued until December 4, 1913—after the Texas City unit had disbanded and reformed at San Diego.

When the unit was functioning at Texas City, more officers and planes joined it. Although the flying field there was adequate when it was selected, two infantry brigades moved into the area and considerably reduced the space available for landings and takeoffs. Three of its sides were obstructed by tents, houses, and wires, and the fourth by Galveston Bay. This ever-present danger, plus high winds, interfered with flying and allowed only the experienced fliers to take to the air.

The conditions didn't prevent record setting, however. On March 28 Lieuts. Milling and Sherman established a new

American two-man duration and distance record by flying the 200 miles from Texas City to San Antonio in four hours and twenty-two minutes. Three days later Lieut. Sherman made the first Army aerial map as he and Milling flew from San Antonio back to Texas City. He made the map in sections for every ten minutes of flight and plotted railroads, bridges, wagon roads, towns, streams, woods, hills, and prairies in great detail. This map was a milestone in aerial navigation because it was the first time one was made from a pilot's point of view for other pilots.

While all this activity had been going on, I had returned to Leavenworth where I received orders to join the 7th Infantry at Fort Crockett near Galveston as an Infantry officer. This meant no official flying duties for me. I was assigned routine camp duties at nearby Fort Crockett and could only watch the planes at Texas City from a distance. My heart ached every time I heard one of the Texas City planes fly overhead. When I could stand it no longer, I went to my commanding officer and asked if I could have time off to keep up my flying proficiency. "Yes, Foulois, you can go over to Texas City," he answered. "But you go over there on your own time—not the government's!"

This kind of retort from a nonflying superior has plagued pilots ever since those early days. Even today I still hear of pilots having to try to maintain their flying skills on holidays and weekends simply because their ground-bound bosses won't give them time off to fly during the week. This shortsightedness, especially in this age of supersonic flying, can and does lead to accidents. Whenever I hear of a pilot with a ground job getting killed on a weekend, I wonder if his superior, possibly jealous of the lad's flight pay, realizes that he may be a contributing factor to his death.

I got in touch with Captain Chandler and he readily agreed to let me fly any time I could get off. This off-duty association with the new planes was valuable because it kept me up to date on what was going on. I met the new pilots and got to know them well. I observed who was having difficulty and who seemed to adjust to the hazards of flying best. I made mental notes of their idiosyncrasies just in case they ever served with me in the future. In addition, I put on my old coveralls and got my hands dirty on the new planes—most of which were *Wright-C*s.

When it became evident that the United States would not be involved in any military operations on the Mexican border, most of the aviation personnel and their equipment at Texas City were transferred to San Diego in June.

I was still at Fort Crockett in August, 1913, when I received telegraphic orders to report to Washington to appear before the Military Affairs Committee of the House "in connection with aeronautical matters." These "aeronautical matters" were hearings on H. R. 5304, a substitute bill for H. R. 28728 which had failed to pass in the previous session.

I arrived in Washington not knowing exactly what was expected of me until I was briefed by Maj. Samuel Reber who was scheduled to become head of the Aeronautical Division. Reber was an old friend and one of the most efficient administrators I have ever known. He had recommended that I be called to testify because of my flying experience "and because you call the shots exactly as you see them," as he told me. I knew of no other way to operate, I told him.

Reber then told me of the criticism that had been leveled against the Signal Corps because of the high number of serious accidents that had occurred in 1913 (seven that year as compared with six in the whole period 1908-1912). "They say these accidents are owing solely to poor administration by the Signal Corps and the War Department," Reber said. "I am therefore recommending that drastic steps be taken to get Army aviation out of the show business and shape it up as a military business. I need your help, Benny, and I think you're the man to tell the Congress. Somehow or other, you've made a reputation with Congressman Hay and he wholeheartedly agrees to your appearing before his committee."

I really got enthusiastic about Reber's plans. He recommended more difficult requirements to become a military aviator, the establishment of an air depot, the activation of a provisional aero squadron of eight planes for tactical service with ground troops, and the permanent establishment of the 1st Aero Squadron at the Signal Corps Aviation School at San Diego. These were exactly the same things I had recommended when I was asked for my comments the year before.

Congressman Hay held his hearings in mid-August and nearly all of the witnesses, including Capt. Billy Mitchell, not yet a flier, recommended against the adoption of the bill that would create a separate air corps until our aeronautic capabilities had advanced to the point where its separation from the Signal Corps could be sustained. It is interesting to note here that Mitchell and his worshipers forgot this episode when Mitchell radically changed his views about a separate air force in later years.

The bill was not reported out of committee this time but there was some benefit derived from the hearings. Senior Signal Corps officers took a new look at their "baby" and began to realize the potential of military aviation. The hearings also gave an excellent opportunity for the Signal Corps to present a compilation of aeronautical progress being made by all the major world nations. The information submitted to the committee included statistics on each country's military planes, airships, and pilots. One item of interest was a comparison of the annual military aviation expenditures of twenty-six governments in the preceding five years. Germany headed the list with $28,000,000 while the United States ranked fourteenth with expenditures totaling only $435,000. Even Mexico, a possible enemy, was spending $400,000 annually on aviation.

I added my two cents to the testimony and got ready to return to Galveston. I dropped by Reber's office to pay my respects before leaving and he told me his troubles. Capt. Charles DeF. Chandler and Lieuts. Lahm, Arnold, and Sherman had just been relieved of aviation duties and returned to troop duty. Lieut. Tom Milling had been sent to France. The school at San Diego needed help because its new commandant, Capt. Arthur S.

Cowan, was not a flier and with all the experienced hands leaving the fold, things were sure to go from bad to worse. Would I consider going back full time to aviation duty and act as a trouble shooter for the commandant at San Diego?

I literally leaped at the chance. Instead of going back to Texas, I proceeded to Fort Leavenworth to pack my household goods, then traveled to San Diego, arriving during the Christmas holidays.

The school at San Diego had had a short but interesting history. Glenn Curtiss had established a school and experiment station at North Island in January 1911. He had partially trained Lieuts. Beck, Walker, and G. E. M. Kelly, but they had been ordered to join me at Fort Sam to fly the Curtiss plane before they had finished the course. Although Curtiss offered to train more pilots on his machines, no more were sent until November, 1912. At that time Lieuts. Harold Geiger, Lewis E. Goodier, Samuel H. McLeary, Lewis H. Brereton, and Joseph D. Park arrived, all of whom had recently been at the Curtiss plant at Hammondsport for instruction. Three Curtiss planes were sent from College Park and on December 8, 1912, the Signal Corps School was officially established. The Army's first flying boat was also shipped to San Diego and flying began. Crashes were regular and frequent.

In March 1913, three more students arrived—Lieuts. Rex Chandler, Joseph E. Carberry, and Walter R. Taliafero. Other students were assigned in the following months. By the end of the year, however, Lieuts. Rex Chandler, Joseph Park, Loren Call, Moss Love, Eric Ellington, and Hugh Kelly had died as a result of crashes.*

Since I had Reber's sanction to get the school into shape, Captain Cowan gave me free rein. The first thing I did during my two weeks' observation was quietly to order two dozen pair of coveralls through the quartermaster at nearby Fort Rosecrans. I drew up two General Orders which Captain Cowan put into effect. General Order No.1 contained a security regulation which stated

* Chandler Field at Essington, Pennsylvania, Park Field at Millington, Tennessee, Love Field, Dallas, Texas, and Ellington Air Force Base, Houston, Texas, were named for these officers.

that "officers, enlisted men, and civilians on duty at this school will, under no circumstances, give out for publication any information concerning the work of the school: Such information as is considered proper for publication will be given out by the commanding officer." General Order No. 2 included a requirement for all personnel connected with the school to wear helmets on all flights, leather jackets when flying over land, and life jackets when flying over water.

There was a very good reason for the first order. Here are my notes. Just as I wrote them in January 1914, before I drew up General Order No. 1:

> ... Another factor that should be considered in the mental fitness of a military aviator is his susceptibility to newspaper praise or criticism, or the desire to obtain newspaper notoriety. This factor has contributed indirectly to the death of a number of our best civilian aviators.
> ... During the early days of aviation, some of the most noted and widely advertised civilian aviators became, through newspaper notoriety, so thoroughly conceited as to think that they knew all there was to know about aviation, that they would absolutely disregard the advice and warnings of their instructors and more experienced aviators, with the result that all of these men have either been killed or have stopped flying as a result of the tremendous strain to which they subjected themselves in their efforts to keep their names on the front pages of newspapers.

The second General Order was just commonsense since most fatal injuries seemed to be caused by skull fractures. In addition, Chandler had drowned when he had crashed with Brereton and Lieut. C. Perry Rich had drowned when he fell into Manila Bay in November 1913.

By the middle of January I had a detailed set of rules drawn up. A full-duty day was prescribed for the student pilots, which included ground-school classes in radio, topography, and mechanics. The latter program was no mere classroom recitation period. I passed out the coveralls and gave the former polo-playing student aviators the same advice that the Wright brothers had given me: “If you want to learn about flying machines, you have to get your hands dirty.”

There was plenty of grumbling, but results were immediately evident. During January we worked with the Navy on experiments in submarine mine detection. A number of cross-country flights were made, most of which were not without incident. But all the incidents added up to valuable experience.

On February 14 Lieut. Townsend F. Dodd and Sgt. Herbert Marcus established an official nonstop American duration and distance record in the Burgess-Wright tractor aircraft when they flew 244.18 miles in four hours, forty-three minutes. Two days later, Lieuts. Carberry and Taliafero set a new official Army altitude record of 8,800 feet.

There was one fatal accident that marred the training program, however. On February 9 Lieut. Henry B. Post crashed in San Diego Bay during an attempt at a solo altitude record in which he reached an unofficial height of 12,140 feet, exceeding by 500 feet Lincoln Beachey's official American record of 11,642 feet. Post had been flying old *No. 10* that had scared Hap Arnold so badly at Fort Riley in 1912 and had been assigned to me at Fort Leavenworth. Post either fell or leaped from the machine at about 300 feet while in a steep dive. The official report cites the probable cause of the accident as “due to pointing the machine down too steeply.” In view of the fact that Hap Arnold had almost been killed in an uncontrollable spin in the same plane, I think the same thing might have happened to Post.*

The plane that had caused Post's death was one of six *Wright-C*s purchased by the government; five of them had killed a total of six men. The sixth was lost in the Philippines when Lieut. Frank Lahm plunged into Manila Bay in September, 1913. Fortunately, Lahm was not injured.

It is important to note here that prior to 1914 the majority of biplanes built in the United States had been of the pusher type with the propeller immediately behind the wings, and constructed by the Wright or Curtiss companies. There was so much disagreement

* Post Field, Fort Sill, Oklahoma, was named in his memory.

among the officer pilots as to which manufacturer built the better plane that two distinct camps of pilots had grown up. Because of the great differences in control mechanisms, pilots checked out on one type did not switch to the other. This competition had become so keen as to be dangerous. Each faction tried to show off the superiority of its machine and training for the benefit of new students, much as college fraternities try to influence prospective candidates. I recommended that the Curtiss and Wright sections of the school be consolidated, and they were after Post's death. I also recommended discontinuance of the use of the *Wright-B* and *Wright-C* machines at San Diego, and that flying be limited to the single Curtiss and the four Burgess-Wright tractors assigned.

In March 1914, relations tensed between the United States and Mexico. To strengthen the nation's air arm at the scene of possible combat operations, the War Department directed the commanding officer of the San Diego flying school to send five qualified aviators, thirty enlisted men, and three Burgess-Wright tractors to Fort Crockett at Galveston. I had just relieved Captain Cowan temporarily as the commanding officer of the school, but with three of the five planes ordered out, I figured my place was where the planes were. I put Lieut. William L. Patterson in charge at San Diego and departed for Texas on April 26, 1914, with my detachment composed of Lieutenants Milling, Taliafero, Carberry, Dodd, and Adna G. Wilde, our medical officer.

This detachment was designated the 1st Company, 1st Aero Squadron, and the men left behind were designated the 2nd Company. We arrived at Fort Crockett just in time to see the last ship loaded with American troops disappear over the horizon. For the next six weeks the 1st Company literally sat on the sea wall at Galveston waiting for whatever the fates decided. We never unpacked our planes and when tempers died down in Mexico, we were finally ordered back to San Diego.

It was our job at San Diego to write the specifications which we hoped would get us a standard machine—one that had the same type of controls so that there wouldn't be any more

"Curtiss" and "Wright" pilots. Standardization, as I look back on it, was our most important problem and one that had to be solved if there was ever going to be a true military air force.

The specifications we agreed upon for training planes prescribed a two-seater tractor biplane with dual controls, maximum speed at least 70 miles per hour, and a minimum speed of not more than 40 miles per hour. We wanted it to carry enough fuel for four hours of flight at its maximum speed. Other desirable features we envisioned were a positive-driven fuel pump, a tachometer, a quickly demountable power plant, the elimination of friction in the controls, and a streamlined design. The machine had to be capable of assembly from packing cases in two hours by four mechanics, while disassembly and packing were to take not more than an hour and a half.

We required certain data from the competitors for the contract, such as stress diagrams and blueprints as well as certificates of engine performance. In order to compete, each machine had to qualify with a four-hour nonstop flight and a 4,000-foot climb in ten minutes while fully loaded. Those that passed these requirements were then graded on a point system on such things as construction, maneuverability, and inherent stability.

Lt. Benjamin Foulois and Orville Wright after the final acceptance tests for the Wright Military Flyer at Fort Myer, Virginia, July 30, 1909. U.S. Air Force photo.

Lt. Benjamin Foulois and Philip D. Parmalee at controls of Wright Type B Flyer, 1911. U.S. Air Force photo.

Enlisted crewmen prepare the Army's first airplane, the Wright Military Flyer designated "Signal Corps No. 1," for flight at College Park, Maryland. The aircraft rests on its launching rail. The tower that suspends the heavy weight that will catapult the aircraft into the air stands behind the craft. U.S. Air Force photo.

The U.S. Army's first plane after it had been equipped with wheels. Pictured are seven enlisted men assigned to the aviation detachment. Location is Ft. Sam Houston, TX in 1910. U.S. Air Force photo.

The 1st Aero Squadron received the first of eight Curtiss JN-2 aircraft at the Signal Corps Aviation School, North Island, California, in June 1915. Capt. Benjamin D. Foulois sits in the cockpit. Vernon L. Burge Collection, Airman Memorial Museum, Silver Springs, Maryland. *Inset picture:* Capt. Benjamin D. Foulois standing in front of a Martin TT training aircraft at the Signal Corps Aviation School, North Island, California, c. 1915. U.S. Air Force photo.

Capt. B.D. Foulois and Lieut. J.E. Carberry picked up by Mexican along road [in wagon] after their "aero plane" had fallen 1500 feet - Mexican-U.S. Campaign after Villa, 1916. U.S. Air Force photo.

Brig Gen. Benjamin D. Foulois and Gen. John J. Pershing, commander of the American Expeditionary Forces (AEF) in France during World War I. U.S. Air Force photo.

Brig. Gen. Benjamin D. Foulois talking with Air Service officers at A. S. Production Center No. 1, Romorantin, France, in 1918. U.S. Air Force photo.

Maj. Gen. Benjamin D. Foulois, Chief of the Air Corps, 1931-1935. U.S. Air Force Collection, National Archives, College Park, Maryland. U.S. Air Force photo.

Maj. Gen. Benjamin D. Foulois explains the air routes covered during the Air Corps airmail operations in 1934. U.S. Air Force photo.

Maj. Gen. Benjamin D. Foulois in the office he preferred above all others. U.S. Air Force photo.

By the time of the competition, twelve manufacturers had declared their intentions to enter. But on October 20, only the Curtiss and Martin companies were ready. The other contestants were barred because they couldn't meet the preliminary requirements.

The competition was called off because there were so few qualified entries. However, we decided that Curtiss and Martin could demonstrate their machines if they wanted to with the understanding that the government was under no obligation to purchase. Only the Curtiss machine met every one of our requirements. Thus, the competition was a disappointment but served to point the way for future military plane procurement for those manufacturers interested. Future planes, of course, met and surpassed these requirements; but it took this first major stretch of their imagination to make them work at it.

From July to December, 1914, very little systematic training or instruction took place because most of our enlisted men were needed in construction work and development of the now officially designated Signal Corps Aviation School. At my suggestion, the school was separated from the 1st Aero Squadron, which I now commanded, to permit us to experiment and practice rather than be tied down to training students. With this clean break, I concentrated on building up an organization that could do a military job if called upon. We obtained eight new planes which built our inventory up to the grand total of twenty-one.

To sharpen the unit, I decided to enter six machines in the Mackay Trophy contest in December 1914. The contest required that military planes make a successful reconnaissance of troops located in a triangular area bounded by the mouth of the San Diego River, La Mesa, and Lower Otay Reservoir. My plan was to have our six planes fly to Los Angeles and start the contest from there. To make a long story short, Capt. Hollis Muller and Lieut. Frederick Gerstner, flying a Curtiss J, landed in the Pacific off Arguello because of high winds where Gerstner drowned attempting to swim to shore; Lieut. Tom

Milling and Capt. Bill Patterson crashed in a forced landing just west of the San Onofre Mountains; Lieuts. Joe Carberry and Arthur Christie were forced down at Pacific Beach with a broken fuel line; and Lieut. Walt Taliafero, with me as his observer, was forced to land when a piece of glass from Taliafero's goggles blew into his eyes. We hit rough ground on landing and went over on our backs.

Only Lieut. Joe Morrow and Capt. Townsend Dodd got their planes to Los Angeles. On the day of the contest Morrow, with Lieut. Ralph Holliday as observer, crash landed near Oceanside. Dodd, with Lieut. Shepler Fitzgerald as observer, made the required flight and thus, unopposed, won the trophy for "the most meritorious flight of the year."

Of the six planes, three were total losses. While these accidents caused us serious concern, the experience was valuable because it showed that, even with experienced pilots and reasonably good planes, adverse circumstances could put a number of machines out of commission in a very short time. This fact emphasized the necessity for having a relatively large number of planes available in case of war, when flying would have to be done in all sorts of weather and over rough terrain, not to mention that extra hazard of enemy ground or air fire.

War was indeed on our minds at that time because fighting had broken out in Europe the previous summer. It was of much interest to note that at the outbreak of hostilities the Allies possessed a grand total of 208 aircraft (Great Britain, 48; France, 136; and Belgium, 24). Germany had 180 planes and was building more when it discovered how useful they really were in aiding ground armies as reconnaissance vehicles.

The first actual airplane bombing by belligerents took place in Mexico in August, 1914, when bombs and hand grenades were dropped by both sides without benefit of a bomb sight. At the same time we decided to test the bomb sight designed by Riley Scott at San Diego, and we began a series of experiments dropping dummy and live explosives, bombs, grenades, and artillery shells. The results with the Scott sight were astonishingly good, so we included bombing instruction at the school from that time on.

Experiments with airplane guns to fire on ground targets were carried out on the East coast and continued by us at San Diego in the fall of 1914. As the year ended, things were looking up. The new year brought a systematic course of instruction, more airplanes, and sufficient civilian instructors to provide a continuous flow of new pilots to the 1st Aero Squadron. Whatever the future held in store for us, we were at least starting to get ready for it.

Chapter Eight

With Pershing in Mexico

The War Department was in the process of estimating the budget for fiscal year 1916 when war broke out in Europe. An Army request for a little more than $1 million for the Aviation section of the Signal Corps was submitted to Secretary of War Lindley M. Garrison, who promptly reduced the estimate by $600,000 and sent an estimate for $400,000 to the Congress. This was done at the same time that Germany appropriated $45 million; Russia, $22.5 million; France, $12.8 million; Austria, $3 million; and Great Britain $1.08 million. Even Italy had appropriated $800,000 for military aviation in its prewar budget.

The Congress was not disturbed by what was happening in Europe. On March 4, 1915, only $300,000 was slated for military aeronautics. When Gen. George P. Scriven, Chief Signal Officer, had appeared before the House Military Affairs Committee, he described an adequate United States air force as consisting of 50 planes divided among four squadrons. Besides the planes, he saw each squadron as best equipped with 18 automobiles and tractors for the transport of spare parts, fuel, and men for each squadron. The four squadrons should have 80 officers and 360 enlisted men. But this was only a dream. The whole Aviation section had only 29 officers, 155 men, and a dozen flyable planes.

It was nothing more than War Department tradition and congressional shortsightedness which was hampering the progress of military aviation. At San Diego we were grateful that our Signal Corps superiors saw the future as we did and that General Scriven had the courage to buck the War Department General Staff on Capitol Hill. But his courage didn't help, and we staggered along as best we could, flying when we had planes ready to fly.

One of the most troublesome provisions of the Act of July 18, 1914, was that specifying that only unmarried lieutenants of the line under thirty years of age could serve in the aviation section. The result was that the section was being filled with young, inexperienced second lieutenants, leaving no one with age and experience to command an aviation organization.

Still another irrational proviso was one that required an officer on aviation duty to return to troop duty as soon as he was promoted to the grade of captain in his basic service. In my case, I was a captain in the Aviation section, but still a first Lieutenant in the Infantry, my basic arm. I was then thirty-six years old, had had seventeen years' Army service, had been on aviation duty longer than any other officer, and had more flying time than any other military pilot. As soon as I would be promoted to captain in the Infantry, according to the law of the land, my flying days would be over and some junior lieutenant would take my place.

In May and June of 1915 we received eight new Curtiss JN2s, which represented the latest in aeronautical developments. They were distinguished by low, rakish lines because of the staggered wings. The fuselage was long, deep, and narrow. They were actually modifications of the British Sopwith and were equipped with instrument panels and a wicker seat for the observer. They *looked* like airplanes. But we were to find that an airplane that looks like an airplane may be something less.

Just after the last of the JN2s arrived, I received orders to take the 1st Aero Squadron to Fort Sill, Oklahoma, for artillery fire adjustment experiments. We arrived there the end of July and tried to get set up. It took us two weeks to make camp, build a mess hall, storehouse, and garage, lay pipe, and install telephone lines. And just as we thought we could get to the business of unpacking our planes, orders were received to send two sections of the unit to Brownsville, Texas, to assist in border patrol work. Four pilots and two planes were despatched by truck with Lieut. J. C. Morrow in charge. As soon as the crews arrived and assembled their JN2s, Morrow sent a message to the Chief Signal Officer in Washington, saying that the Brownsville field assigned was too small for the slow-climbing JN2s. He asked for permission to use another field, rougher than the one assigned, and requested that two Martin tractors be sent for his use. A curt reply came back immediately:

> GO INTO CAMP AT OR NEAR POST. PREPARE FIELD WHICH YOU REPORT TOO ROUGH SO THAT MACHINES CAN START AND LAND. YOU ARE EQUIPPED WITH SERVICE MACHINES AND NONE OTHERS WILL BE FURNISHED. IF YOU CANNOT MEET THE INCIDENTS OF ACTIVE SERVICE YOU WILL BE SUPERSEDED.

With this kind of answer Morrow did the best he could with what he had. But he couldn't overcome the inherent problems that had been built into the JN2. On August 26 Lieut. B. Q. Jones, with Morrow as observer, took off from Brownsville to look for bandits reported operating in the area around Santa Maria, Texas. Jones had a difficult time controlling the plane. He had the same difficulty several days later. Lieut. H. W. Harms aborted a bandit-hunting mission because he said he could not control the plane in rough air.

On September 5 Morrow, with Priv. Adam Khuen-Kryk as passenger, took off and climbed slowly straight ahead. Morrow leveled off at 200 feet and began a shallow left turn. The plane wobbled, stalled, and went into a nose dive. Although he regained partial control, it was not soon enough, and the JN2 smashed into the ground at a 45-degree angle. Morrow was injured and was replaced by Capt. Townsend Dodd who was ordered from San Diego.

In the meantime, B. Q. Jones was left in charge at Brownsville. He immediately rendered a report stating that the JN2 was entirely unsuitable for military purposes and unnecessarily endangered the lives of the men required to fly it.

While the detachment at Brownsville was having its troubles with the JN2s, I was also having mine. We made many attempts to perform reconnaissance missions with artillery troops, but constant engine and air-frame troubles prevented us from being very useful. We had received twelve engines before leaving San Diego, but six of them were immediately condemned and one required an overhaul. Although I asked the Curtiss people to rush me engines and plane parts, nothing happened. Within two weeks all the planes were grounded for lack of spare parts. To make matters worse, Lieut. R. B. Sutton had an accident that was fatal to his passenger, Capt. G. H. Knox, Q.M.C.

After this accident I called a meeting of my pilots to discuss the JN2. Ten of the twelve pilots agreed that it was unsafe and should not be flown, especially with a passenger, because of its low power, cheap and faulty construction, lack of stability, and unduly sensitive rudder controls. Tom Milling and I did not wholeheartedly concur, but since the planes were in such poor shape, I suspended flying for about six weeks while we checked every nut and bolt on each machine and I was satisfied that we could resume our work with the Artillery.

In the meantime, Glenn Curtiss, worried about the reputation of his planes, promised to send new upper wings, stabilizing fins, rudders, and new engines to remodel completely the JN2s and remove the reported defects. He kept his word, and the planes at Fort Sill and Brownsville were modified into new JN3s.

Prior to my leaving San Diego, General Scriven had approved the construction of quarters, a garage, a stable, and a machine shop at Fort Sam Houston on a site I had selected. This was to be the permanent home of the 1st Aero Squadron as soon as we completed our work at Fort Sill. Accordingly, on November 19, 1915, the first squadron cross-country flight in the history of the United States began a distance of 439 miles.

I appointed Master Sgt. Stephen J. Idzorek and Pvt. First Class Westmark as the advance ground echelon. They would precede us and select the best landing field at Wichita Falls, our first stop. They left on November 17 in the squadron automobile. Enlisted personnel and trucks moved by both road and rail, and our route was to follow the railroad so we would have all of our equipment reasonably close to us if we had any forced landings. Besides, following the railroad in the air was a foolproof way of getting to our destination since we had no maps except geological survey maps which were thirty years old.

On November 19 my five pilots and I took our six new JN3s off at one-minute intervals and headed for Wichita Falls. The planes and the squadron train arrived together and created quite a stir among the citizenry. The crowd got so pushy that I had to require each pilot to guard his plane lest souvenir hunters cart it off piece by piece. From then on I had Sergeant Idzorek arrange for police protection at all stops.

We spent the next days hopping to San Antonio via Fort Worth, Waco, and Austin. The only mix-up came when Lieut. Joe Carberry was leading the formation between Waco and Austin. A strong wind blew him off course, and we followed him. After discovering his error, each of us landed at various places, asked directions, and then flew on our own into Austin.

We pilots felt good about this new aviation "first" after arriving in San Antonio, but what was actually more significant was what we had learned about the mobile equipment in our ground echelon. Our eight trucks, with loads of one to one-and-one-half tons, were subjected to a hard test over rough roads and long stretches of deep sand. Some broke through the flimsy wooden bridges they had to cross but I had anticipated this difficulty and sent wrecking equipment in each truck. I had equipped our best mechanics with motorcycles and portable tool kits. These men rode as fast as safety permitted to the next landing field after we took off and met us there to check our planes and engines before the trucks arrived. Both the cyclists and the truck crews worked long hours; they rode all day, worked late at night, and averaged about one meal a day that week. It was their efforts that made the flight a success.

By the end of the year 1915 the 1st Aero Squadron, now fifteen pilots strong, was firmly settled at Fort Sam Houston, Texas. In addition to our six JN3s, we acquired three Burgess planes to bring us up to strength.

On the night of March 9, 1916, Francisco "Pancho" Villa, with a force of from 500 to 1,000 men, crossed the international border and raided Columbus, New Mexico, killing 17 Americans. Considerable property was destroyed before units of the 13th Cavalry drove off the marauders.

The following day President Wilson ordered Brig. Gen. John J. Pershing into Mexico to assist the "official" Mexican government in taking Villa. No better man could have been chosen for the job. Having known and worked with him in the Philippines in 1902 when he was a captain, I had the most profound respect for him as a man and as a leader of men. There was something about his bearing and his personality that attracted attention as soon as he entered a room. There was only one other military man I have ever known who possessed the same personal magnetism that silenced conversations and caused other men to stop, listen, and be impressed. That man was Douglas MacArthur. To all of us who knew both these men as junior officers we would have readily predicted their eventual rise to military greatness.

Pershing was not only the best man for the job, he was also near the scene of Villa's raid. He was in command of the garrison at El Paso, 50 miles from Columbus. This, of course, was the real reason he was selected to chase Villa down. It was to be this "punitive expedition" that would put him in the national spotlight and mark him in the public eye as the proper man for the job when our armies had to go to Europe. When the whole fracas was ended, historians would record the campaign a success in spite of the fact that Villa was never captured and American troops withdrew after eleven months of frustration and disappointment.

There is no doubt that the expedition to chase Villa was hastily organized, but the strategy was simple. The basic plan was to trap Villa between two fast-moving cavalry columns. One column was to head south from Culberson's Ranch and the other was to parallel it, starting from a point east of Columbus. The order from the Secretary of War Newton Baker stated that Pershing was to pursue the Mexican band and that "these troops will be withdrawn to American territory as soon as the de-facto government of Mexico is able to relieve them of this work. In any event, the work of these troops will be regarded as finished as soon as Villa's band or bands are known to be broken up." Secretary of War Baker also instructed Pershing "to make all practical use of the aeroplanes at San Antonio, Texas, for observation" and to "telegraph for whatever reinforcements or material you need."

Three days after the expedition was ordered, the Army's deputy Chief of Staff wrote a memorandum to Secretary Baker, assessing the situation as he saw it:

> They [Pershing's forces] will soon be beyond assured communication with their home government. They will be in hostile country surrounded by enemies, and they will do as soldiers in the circumstances must do. If they think that Villa and his band are in a certain town and refuse to surrender, they must attack that town. If Villa retreats, they must follow him; if he breaks up into small bands, our people must more or less scatter in order to find him.

No truer assessment could be given. A week after the raid on Columbus, Pershing ordered his two regiments of cavalry across the border in "hot pursuit." Before the chase was over, 10,000 American troops were eventually called upon to track down one guerrilla leader and a thousand of his followers. More than 75,000 National Guardsmen were called into Federal service. The net result was not to be a resounding victory but an exercise that was to point up serious flaws in the structure of the United States military establishment—including its active air arm consisting of the eight training planes I had under my command in San Antonio. The newspaper accounts of Villa's raid on Columbus had been harrowing. As I read of Villa's attack and subsequent withdrawal across the border, I realized for the first time in my life that I was visualizing the raid from an airman's point of view. In my mind's eye I saw the locale, the raid, the defense of the town by the 13th Cavalry, and the scattering of Villa's guerrillas afterward as though I were hovering above the scene. In short, I realized that I was thinking about a ground battle from the vantage point of the sky. Although I had forced myself to think this way previously, I found that I was doing it subconsciously now. It is this third-dimensional point of view of ground events that sets the airman apart from his earth-bound colleagues. Experienced pilots always have it. Only rarely is anyone whose life is tied to the ground so gifted that he can truly assess military situations with the omniscience of the flier. This difference has long separated the men of the air from their counterparts in the Army and Navy.

While I wondered what was going to happen, I mentally planned what I would do if the 1st Aero Squadron were ordered to help Pershing. During the three days after Villa's raid I made it a point to talk to all of my men individually to find out what problems they were having both on and off the job. I checked over the eight planes, ten trucks, and one automobile assigned. I looked at maps of Mexico and memorized names of towns, rivers, mountains, their elevations and distances between major points. I reviewed the airplane maintenance problems we had and inventoried the stocks of spare parts, gasoline, and personal equipment for the men. The answers to my self-asked questions were disturbing. We might be able to last thirty days, I told myself, but after that we would probably be in deep trouble for parts. The sand and heat in the daytime and the cold of Mexican nights would probably take enough of a toll of parts, planes, and men. To add a requirement to fly many missions in support of ground troops would put a severe strain on the squadron. I gritted my teeth, drove my men to distraction with my questions about their personal readiness to go into the field, and watched the newspapers.

Prior to Villa's raid on Columbus, New Mexico, the headquarters Aviation section of the Signal Corps in

Washington was under investigation by the Congress and the War Department. The probe resulted in the relief of the chief of the section, Col. Samuel Reber, Signal Corps, and the temporary assignment of Capt. William Mitchell, Signal Corps, to Reber's post. This change in the headquarters Aviation section command was, in my opinion, most unfortunate, especially at a time when we were faced with a possible war with Mexico. I had personally known Colonel Reber ever since the beginning of our military aviation in 1908. He was one of the most experienced officers in the Signal Corps, an efficient executive and administrative officer, and thoroughly acquainted with the problems of organization, training, and operations encountered by the Signal Corps during 1908 to 1916.

I had also known Billy Mitchell ever since I started flying in 1908, and to the best of my recollection he had never taken an active personal interest in anything having to do with the Aviation section of the Signal Corps until 1916, when he was assigned to replace Reber.

I soon learned that Reber's relief from aviation duty was chiefly owing to political pressure on the War Department, and that the War Department had insisted a Signal Corps officer with General Staff experience be assigned as his replacement. Hence, the assignment of Mitchell, a qualified War Department General Staff officer, but entirely inexperienced in solving the practical problems incident to flying,

In addition I also learned that one of the basic reasons for the assignment of Mitchell was primarily to instill old-fashioned military discipline among the so-called prima-donna pilots then on active duty. This was to me an extraordinary and amusing reason, and as it turned out, a supreme irony which almost wrecked military aviation in this country.

I was older than Mitchell. He and I began our military service as enlisted men in the Spanish-American War (1898); however, he served only seven days as an enlisted man in the 1st Wisconsin Infantry. His father, then a United States senator, made him accept a discharge and a commission as a second lieutenant at the age of eighteen years!

The noncommissioned officers of the old, regular Army were then, the basic disciplinarians in the Army, and had Mitchell been subjected to the same hard-boiled type of military discipline which I had experienced as an enlisted man, it is my sincere belief that he would never have terminated his active-duty service with a military court-martial conviction.

As I expected, when Mitchell took over Colonel Reber's job, he called in several officers from the Signal Corps Aviation School at San Diego, California, and assigned them to duty in his office. All of these young officers were good men, but totally lacking in flying experience.

At noon on March 12, 1916, a Saturday, orders arrived directing the 1st Aero Squadron to proceed by rail to Columbus, New Mexico. I reacted automatically, as if I had been practicing, and I guess I had, mentally. Within four hours the squadron's complete equipment was packed and ready to go. The next morning we flew the planes to the Fort Sam drill grounds where they were disassembled and loaded on flatcars. Shortly after midnight we left San Antonio and arrived at Columbus about thirty hours later.

The orders given to me were merely "to proceed." The basic assumption was that our planes were to be used primarily for observation, both of enemy forces and advanced parties of the expedition, for the carrying of mail and dispatches, and for aerial photography. However, before we left San Antonio I had requested that we be provided with a Scott bomb sight and bombs, and that we be permitted to equip our planes with machine guns, as I had heard the Germans and the French were doing in Europe. I never received sight, bombs, or guns. The reason was simple. They weren't to be had.

Moving with the 11 officers were 82 enlisted men and one civilian mechanic. At El Paso we were joined by a medical officer and three hospital corpsmen. In the next three months I had more officers and men assigned. However, none of the additional men had much experience on aviation duty, and I found it difficult to try to run combat missions and train newcomers at the same time.

We arrived at Columbus on March 15, 1916, and immediately began unloading and assembling our planes. When the first one was ready the next day, Capt. Townsend F. Dodd, my deputy, and I made the first reconnaissance flight 30 miles into Mexico—the first aerial reconnaissance by United States military aircraft ever made over foreign territory. Although we had a negative report to make to General Pershing, this fact alone was significant for it meant that there were no Mexican rebels within a day's march of the head or flanks of his infantry and cavalry columns.

On the morning of March 19 I received telegraphic orders to move the squadron to Casas Grandes, a small town about 125 miles south of the border, for immediate service. By five thirty that afternoon all eight planes were in the air headed for Ascension, about 60 miles to the south, where we planned to stay overnight. I was proud of the fact that we had managed to get all our planes air-borne, because it was the first time that we had a combat order to obey and we were 100 per cent ready. I was proud too soon. Ten minutes after takeoff one plane dropped back and was forced to return to Columbus. By the time we reached Ascension, the sun had gone down and I could see only four planes in my formation. Fortunately, the men on the ground had built fires around the landing area they had chosen for us. By the time the fourth plane landed, it was dark.

One plane, piloted by Lieut. H. A. Dargue, who was trailing the four leading planes, had failed to see us land and continued a few miles farther. He landed safely in an open valley with nothing but luck to guide him, remained with the plane all night, and then rejoined us the next morning at Ascension. Another plane, piloted by Lieut. R. H. Willis, also continued south without seeing us land and after a vain search for our landing fires landed in the open desert, 40 miles beyond our troops. He damaged the landing gear and could not get off the next day. He was in hostile territory so he left his ship where it was and started to hike toward Casas Grandes. He hiked all that night, hid out on a mountaintop the next day, and rejoined us before dawn of the second day. I sent a truck with spare parts and an armed guard to his plane, but the Mexicans had gotten there first. The plane was literally chopped up in pieces and everything was destroyed but the engine.

Lieut. E.S. Gorrell, in a third plane, had also proceeded beyond Ascension in the darkness. When he was unable to locate the landing fires, he turned northward hoping to find an American camp or make it back across the border. His fuel exhausted, he landed near a lake 40 miles east of the line of march and about 50 miles south of the border. Although he landed three hours after dark, he lucked out and did not damage the plane.

Meanwhile, we who had landed at Ascension on the night of the nineteenth left the next morning for Casas Grandes where, to our surprise, we found no American troops and no possible area near the town where we could land. After reconnoitering for miles in every direction, we finally landed about 10 miles to the northwest. Headquarters, I found out later, were located near the Mormon colony of Colonia Dublan instead, a difference of another 15 miles. But this error turned out to be a lucky one. I found out that my request for bonfires to be lighted around the four corners of a suitable landing field in the event of a night landing had been received and dutifully carried out. However, our opinion of what was suitable was entirely different from that of the officers with the advancing American column. The field they had chosen was surrounded on three sides by a river and tall cottonwood trees. The field was dry marsh, and was covered with so many high clumps of grass that an automobile could not have traversed it. The fires had been built under the trees and were used by the soldiers for their evening meals. If we had all made that flight and attempted landings on their field, the entire combat air force of the United States would have been wiped out in a few minutes.

When I reported to the division commander next day, I was

instructed to make a reconnaissance flight south toward Cumbre Pass in the heart of the Sierra Madres in an effort to locate rebel troops moving south toward Lake Babicora. About noon Dodd and I took off and headed south. I was worried because the mountains loomed ahead much higher than I had ever been in an airplane, but orders were orders. All we could do was try our best.

About 25 miles out I noticed the ground getting closer and closer. The plane began to buck and shake violently in ever-increasing turbulent air. We had the throttle on as far as it would go, and the engine was straining valiantly at its limits. Ahead, the ground rose abruptly above us, and I knew that much as Dodd and I wanted to complete our mission, we had reached the absolute ceiling of our plane. We had no choice but to return to headquarters and report our failure.

While I was fretting about our inability to carry out orders after landing, I looked up just in time to see Lieut. Tom Bowen get caught in a vicious whirlwind just as he touched down, and end up in a pile of splinters and cloth. He suffered a broken nose, cuts, and bruises; the plane was a total wreck. We now had only six planes left, a 25 per cent loss during the first two days of operations.

The next day Dodd and I had better luck when we were ordered to locate American troops in the Galeana Valley. We found them, landed, took their request for supplies, and returned to Colonia Dublan. As a result, six of the squadron trucks were loaded and sent to the advancing column.

Our meager success was short-lived, however. On the twenty-second, two planes, piloted by Dodd and Lieut. Carleton G. Chapman tried again to fly toward the Sierra Madres to locate American troops moving south along the Mexican Northwestern Railroad. Again the violent air currents and the inability of the engines to pull the planes over the hills forced them to return to Colonia Dublan.

That night I wrote a long memo to General Pershing, and told him that the JN planes I had "were not capable of meeting the present military service conditions" and pleaded for ten newer planes, a spare engine for each two planes, and spare parts. I further requested "that this order be placed by telegraph and immediate delivery of all equipment, by express, be specified."

I knew I was optimistic in thinking I would get even one of the planes I wanted, but I was duty bound to ask for them. In the meantime, we would do what we could within the limitations of our equipment.

The day after I despatched the message to Pershing, three of our planes piloted by Lieuts. C. G. Chapman, J. E. Carberry, and A. R. Christie located the troops in the Galeana Valley and landed. However, they could not return for three days because of high winds, dust, and snowstorms.*

Between March 27 and 31 we managed to make about twenty flights carrying mail and despatches, during which we all encountered severe rain, hail, and snowstorms. By the end of our first ten days of operations it was obvious that our six planes were incapable of fully performing the task assigned. Their low-powered engines and limited climbing ability with the necessary military load made it impossible to operate them safely in the vicinity of the mountains. In addition, the dry atmosphere was hard on the wooden propellers, causing them to warp and the laminations to come apart.

During the entire Mexican campaign one of the most frustrating conditions under which we operated was the refusal of the Carranza government to let us use the railroads for transportation or visit the villages for supplies. The available motor-truck and pack-train supply was far from adequate, to put it mildly. The farther our troops proceeded south, the more their hardships increased, To add to these difficulties the Carranza forces, whom we thought we were helping chase Villa, were openly hostile to us.

* The dust in the air was so thick that the snow was actually brown by the time it hit the ground.

When the supply shortage began to get critical, I was asked to fly down to the city of Chihuahua to contact the American Consul there, Marion H. Letcher, to see what could be done about getting critically needed medicine and food items. The town was held by the allegedly friendly forces of Carranza, but I was suspicious. The reports I had seen from the various commanders trying to locate Villa did not indicate any friendship because they had been fired upon by Carranzistas.

I decided to send two planes with pilots and observers and duplicate messages for Mr. Letcher. One plane was to land north of the city and the other on the south side. The observers were to walk to the city from opposite directions while the pilots would protect their machines, and fly them out, if necessary, to prevent their capture. Herb Dargue and I took off in one plane, while Joe Carberry and Townsend Dodd departed in the other. Carberry and Dodd landed on the north side of Chihuahua without incident. Dodd commandeered a carriage and was driven directly to the consulate. Letcher contacted a few merchants and such supplies as were available were immediately purchased and arrangements made to have them shipped by train later that day to the town of San Antonio.

While Dodd was having success, I was not. A number of townspeople had seen us circling south of the city, and came running toward the field we selected. Four Mexican rurales were waving rifles excitedly and didn't look as pleased to see us as we had hoped. When Dargue got the airplane stopped, I yelled to him to take off immediately and fly north to join Carberry and that I would try to meet him there later. He nodded, gave the engine full throttle, and zoomed over the heads of the approaching rurales, leaving me sputtering and splattered with dust.

I immediately started walking briskly toward the city, and tried to ignore the group shouting and shaking their fists at the departing plane. Four shots rang out, but Dargue got away. I shouted at the crowd to divert them. The rurales wheeled and leveled their rifles at me. I was defenseless except for a Colt

.45, which was no match against four Winchester rifles. There was nothing I could do but put my hands up—and pray. I did both.

As fast as it takes me to tell it, I was a prisoner and found myself being pushed and prodded toward the city jail. Although I did not speak fluent Spanish, I knew enough to realize I was in trouble. There was hate in the faces of the hundred or so men and boys who were herding me down the street yelling *"¡Mata el gringo!"* ("Kill the Yankee!"). They crowded and pushed me, trying to provoke me and give them an excuse to do me in. Never in my life have I ever fought an impulse to grab something and hit back so much; but I managed to keep my temper, hoping we'd get to the jail soon. As the trigger-happy crowd pushed me through the streets, I heard a voice shout in English, "Do you need any help, Captain?"

I looked along the sidewalk and saw a man with Caucasian features standing among the Mexicans. "Yes," I shouted back. "Go get the American Consul!" I was prodded with several rifle barrels at that instant and could not tell whether he heard me.

When we arrived at the jail, I was thrust through the doorway and into a cell. An iron door clanged shut behind me, and I became the first United States aviator ever to become a prisoner of war.

Dodd, in the meantime, was having all kinds of success. While the supplies were being loaded on a train, Mr. Letcher, the American Consul, took him to see the civilian governor of the state of Chihuahua who turned out to be an old schoolmate of Dodd's at the University of Illinois. While they were chatting amiably, word came that an American aviator was in the city jail.

"That must be Foulois," Dodd told his hosts. "How do we get him out?"

The civilian governor merely shrugged. "I have no jurisdiction over the police," he said. "You must see the military governor."

In the meantime, I had begun to argue for my release with

the jail warden. In my most polite "Tex-Mex" Spanish, I tried to explain that I was not an enemy of the Mexican people. In his best English, the warden tried to tell me that I was guilty of a crime until I was proven innocent. After much polite argument, he finally agreed to send a messenger to General Gutierrez, the military governor. However, he noted, it would soon be time for siesta, and the messenger might have to wait several hours before he could gain entrance to the headquarters.

I was lucky. It was not yet siesta time. A Colonel Miranda, the general's chief of staff, showed up, took me in custody, and we marched several blocks to the headquarters. General Gutierrez was affable and agreed that I should not be detained. I told him about the two planes north of the city and asked for guards to keep them from being harmed. Again he was agreeable. I asked if I might visit the planes to reassure my men, and we were soon on our way. As I looked back over my shoulder, I saw the general stretch and yawn. It was obviously siesta time now.

Upon arrival at the landing field north of town, only Herb Dargue was there, and had a harrowing tale to relate. He had joined Carberry after letting me out, but his arrival had drawn a huge crowd of Carranzistas who proceeded to gather around both machines, making insulting remarks. They burned holes in the fabric with cigarettes which made Dargue and Carberry slightly uneasy. When they tried to stop this, several of the mob slashed at the cloth with their machetes. Young boys began to swarm all over the planes, loosening nuts and turnbuckles.

The two pilots felt their only defense was to make what Dargue called a "strategic retreat." They jumped into their planes, fired up the engines, and headed for the wild blue yonder. Carberry got off all right but dusted the mob so thoroughly with his propeller blast that they chased after the other plane, throwing rocks as Dargue lined up for takeoff. Dargue's plane was just lifting off the ground when the entire top section of the fuselage behind the cockpit flew off and struck the vertical stabilizer. Dargue immediately chopped the throttle and landed straight ahead. When I arrived with the guards, he was doing his best to hold off the angry mob with his wits and bare hands and a loud voice. The guards immediately took charge and quieted the crowd down.

Carberry landed in a field beside the smelters of the American Smelter and Refining Company six miles away, and returned later that afternoon. When the mob was finally dispersed, Carberry, Dargue, and the Mexican guards repaired the damage to both planes as best they could. I returned to the city, met Dodd, and the two of us finished our business of getting supplies loaded. That evening the local newspapers came out with big headlines—"Enthusiastic Reception to the American Aviators." It had been enthusiastic, all right, but not in a way that made me very appreciative. The four of us could have been killed by that kind of enthusiasm.

We stayed overnight at the American consulate. Next morning, before sunup, we took off to return to San Geronimo. Dargue and I got off first, and headed out on course. For some reason they could never make me understand, Carberry and Dodd decided to fly over the city for a last look. As they did so, several hundred rurales saluted them from the rooftops by firing their rifles. As Dodd told me later, the only uncomfortable part about this demonstration of affection was that the rifles were loaded with ball cartridges and aimed at his machine. Thanks to the poor marksmanship program of the Carranza army, no hits were scored.

After this mission, we moved the squadron first to San Antonio and then to Satevo, meanwhile keeping our message-carrying routine going. But American troops were not what you would call successful. Villa had not yet been found, and the deeper we went into Mexico, the more hostility we met from Villa sympathizers and Carranzistas alike. On April 12 a small cavalry squadron under command of Maj. Frank Tompkins had a running battle with Carranzistas near Parral. Forty of Carranza's men were killed and many more wounded. Two Americans died and six were wounded. The supreme irony of the whole encounter was that both the Carranza and the American forces were supposed to be going after the same enemy—Villa. The unnecessary deaths of 42 men pointed up the ridiculous position the American forces found themselves in trying to locate and punish a Mexican renegade.

On April 14, 1916, Herb Dargue and Ed Gorrell made a reconnaissance flight from Columbus in an attempt to locate a large Carranza force reported in the area. They never found it, but their flight set an unofficial American distance record for two persons of 315 miles. The same day Lieut. I. A. Rader damaged his plane landing on rough terrain near Ojito, 100 miles from the nearest base. He had to abandon it, leaving us by this time with only three planes. Three days later Dargue and Willis made a forced landing while on a photographic mission from San Antonio, Mexico, to Chihuahua, and demolished the plane. Willis, who was pinned under the wreckage, suffered a severe scalp wound and badly bruised legs. The two of them burned what was left of the plane, starting a huge forest fire in the process, and began to hike to San Antonio, their takeoff point. Two days and nights later they limped into camp and told their story. A month later doctors discovered that Willis had a broken ankle. His reply to that was, "Well, I guess that's why it always felt better when I sat down than when I stood up."

With the force down to 25 per cent of our original strength now, I knew that the days of the 1st Aero Squadron were numbered if we didn't get new planes. The bad news about our planes had gotten to Washington, and a special appropriation of $500,000 was asked of the Congress. "We now have actual field operations going on," Secretary Baker told the uninformed congressmen. "The aeroplane service is, of course, the scouting service, and I would like, if possible, to buy some additional aeroplanes at once, taking advantage of the experience that manufacturers in this country have had from the European experience, and send them down there and get the value of testing them under actual field conditions."

Surprisingly, the Secretary got what he asked for. Twelve new Curtiss R-2 planes were ordered which were to be equipped with Lewis machine guns, automatic cameras, incendiary and demolition bombs, and radios. This was good news, and on April 20 I received orders to return what was left of the squadron to Columbus to pick up the new planes. When we arrived, we parked our two remaining tired and battered Jennies on the side of the field and later put a match to them. I didn't want to take any chances that somebody would order us to keep on flying them.

Instead of new Curtiss R-2s awaiting us, however, there were four Curtiss N8s, which were duplicates of the JN4s that Curtiss was making for overseas delivery. Although they had slightly more powerful engines than our original JNs, I immediately declared them unfit for Mexican service after giving all four of them exhaustive test flights.

My unhappiness with the N8s got to the press, but the reporters had not gotten their stories from me. The only action I had taken was to make an official report to General Pershing on the unsatisfactory performance of the N8s. A reporter from the *EI Paso Herald* had either gotten hold of the report or had speculated correctly that all was not well. I had to defend myself to General Pershing and report that the only action I had taken was official and was contained in the proceedings of the board that had backed up my recommendations condemning the N8s. Pershing, in endorsing my reply to Washington, said that he would have no quarrel even if my officers had expressed public disapproval of the old wrecks we had to fly. "They have already too often risked their lives in old and often useless machines which they have patched up and worked over in an effort to do their share of the duty this expedition has been called on to perform," he wrote. He then said some kind words about our fearlessness and devotion to duty, and suggested that our recommendations should be given a great deal of weight because we were in the best position to judge the planes' performance.

The Curtiss R-2s arrived eventually, but they were not the "dream planes" we had been told to expect. For the next three months we had constant engine and construction troubles. Every plane required alterations and replacement of vital parts. The biggest problem turned out to be the propellers which had been manufactured all over the States and sent to us for testing. To do this we had to construct an engine and propeller test stand. Practically all the propellers tested were defective because of the dry climate. As a result we never again were able to perform useful field service with the Pershing forces. However, we did manage to get a half-dozen planes in the air on August 22, 1916, and give General Pershing the first aerial review ever held by a United States air unit.

By the end of August negotiations between the United States and the Carranza government had been going on for four months to settle the strange questions which had arisen. Border raids by Mexican bandits continued while the Carranza government insisted that it could control Villa without American intervention. Pershing, meanwhile, withdrew his forces to Colonia Dublan and waited for further orders. We continued our efforts to get our planes operational at Columbus, where we found ourselves turning into a service test organization rather than a fighting air force.

In September, 1916, I was ordered to Washington for temporary duty in the office of the Chief Signal Officer. Two months later I was back at Fort Sam Houston as Aviation Officer for the Southern Department. In January, 1917, American forces were ordered out of Mexico and the last man crossed the border on February 5.

Historians have concluded that President Woodrow Wilson's overall policy regarding Mexico was both ineffective and unsuccessful. His military policy must be considered likewise. His dispatching of the punitive expedition without adequate backing, while assigning to it a most difficult mission, was an improper application of military persuasive force.

Pershing had almost limitless restrictions imposed upon him. He had to respect Mexican sovereignty while violating Mexican territory. He had to avoid entering towns, even though Villa's bandits might be seeking refuge there. He had to respect and cooperate with Carranza's forces, even though they were equally antagonistic. He had to try to capture Pancho Villa in an area where Villa was considered a Robin Hood to the populace. He had to supply his forces by motor vehicle while the use of a railroad belonging to the government he was trying to help was denied him. He had been placed in the awkward position of a lion tamer sent into a cage with two hungry lions inside. His orders were to tame one of them but under no circumstances could be do anything about the other one.

Although Pershing failed to capture Villa, the activities of the American troops in Mexico and along the border were not wasted effort. The dispersal of his band by the penetration of American forces deep into Villa's territory put an end to serious border incidents. But more important from a military point of view was the intensive training in the field received by both the regular Army and the National Guard on the border and in Mexico. Many defects in the military establishment, especially in the National Guard, were uncovered.

From the point of view of the airman, I consider the experience of our eight-plane air force to have been a vital milestone in the development of military aviation in this country. The machines were inadequate for the task assigned. Not only were they inadequate, they were downright dangerous to fly because of their age. Yet, we did a great amount of scouting over country in which cavalry and infantry could not operate, and we began and maintained the first regular aerial mail route for the United States and delivered thousands of letters to and from Pershing's troops.

The work of the 1st Aero Squadron proved beyond dispute to the most hardened former soldier congressman that aviation was no longer experimental or freakish. There is no doubt in my mind that the support later given to the Air Service by both General Pershing and Secretary Baker had its roots in the performance and determination of my splendid crew. Proof of the awakening interest in military aviation came in the form of the first appreciable aviation appropriation ever made in the United States—$13,281,666—which was approved by the Congress on August 29, 1916.

My reports had been widely circulated in the War Department, and my own concern and frustration because of the lack of organization of Army Aviation had been transmitted to Secretary Baker. In hearings on Capitol Hill he made the following statement:

> The men in the Aviation Corps have been almost exclusively comparatively young men, very young men, indeed, and they have been engaged in an art desperate, daredevil, hazardous indeed, so that they have an attitude toward life and toward themselves that men have who are engaged in an especially hazardous service. Being fliers, they have had rather a disposition to chafe at the restraint and discipline which was made for more normal kinds of service, feeling that they were not adapted to the regulations and restrictions of men who were not engaged in so unusual an occupation. In other words, they had an impatience at being controlled by men who did not themselves know the business in which they were engaged. I do not want to be understood to criticize these young men. They are pioneering for the army and the United States and their exploits are superb. The net result of it all is that I am going to reorganize the entire Aviation Section.

I take those words to be a compliment to the 1st Aero Squadron. They marked the beginning of my crusade for American air power should we be called upon to help fight a war in Europe that was already two years old.

Chapter Nine

Planning to Fight the First Air War

My only contact with Washington during the spring and summer of 1916 while with the 1st Aero Squadron had been confined to correspondence with the staff of the Aviation Section of the Signal Office, much of which had been answering their long-distance criticism of my handling of the operations. However, in May, 1916, Col. George Squier was assigned to the Aviation Section as chief (Mitchell did not move up), and I began to receive constructive telegrams and letters which indicated his personal interest and concern with my logistical problems. One of his first acts was to ask me to prepare an equipment manual for aviation units which called for a complete inventory of every item of matériel then in use by the 1st Aero Squadron.

Encouraged by the constructive help and sympathy I was getting from Squier, I decided to go to Washington and learn what I could from him of his recent activities in Europe where he had been on duty as a military observer. I was granted a three-month leave of absence by General Pershing, and proceeded to Washington in September. When I arrived, I received a chilly reception from Mitchell and his staff of young pilots, but not from Squier whom I had known since Fort Myer days in 1908 and 1909.

Squier and I held our meetings in the Army-Navy Club in downtown Washington, because he did not trust Mitchell and did not want him to know what we were discussing. During our many long discussions, Squier thoroughly briefed me on the development of European aviation and I briefed him on the pitiful conditions of the 1st Aero Squadron and my experiences in Mexico. He finally asked me to give up my leave and go to work on temporary duty in his office to make plans for American military aviation in the war we both knew was inevitable.

Secretary Baker had announced to the Congress that he had a plan to "reorganize the entire Aviation Section" in view of the fact that the Congress had given us the first appreciable appropriation (more than $13 million) for military aviation that had ever been authorized. My job was to plan around this $13 million for the organization, training, and operation of the new air units. One of my first tasks was to obtain a suitable flying field, and I could not envision it being anywhere else but in San Antonio. However, it would be impossible, in my view, to use the old Remount Station on the Austin road. Another site had to be found.

I returned to Fort Sam Houston and began an immediate reconnaissance of the San Antonio area. I selected a site which later was named Kelly Field. However, when I tried to set the wheels in motion to obtain the land, I learned that no provision had been made in Secretary Baker's plan for the purchase of land or the construction of facilities in the Western or Southern portions of the United States. I found out later that Billy Mitchell was the one who had a big part in the writing of the congressional legislation that had gotten the $13 million, and had had language in the bill that would allow money to be spent only in the Northern and Eastern sections of the country—where the flying weather was relatively poor compared to the rest of the country. I was amazed at such stupidity, and again asked for a leave of absence in December, 1916, to go to Washington to right the wrong that Mitchell had allowed to be committed. Through the efforts of Sen. Morris Sheppard of Texas, then chairman of the Senate Military Affairs Committee, the 1918 appropriation was finally amended to provide funds, not only for the purchase of Kelly Field, but also for other land in the Western and Southern part of the country. Having brought this glaring deficiency to the attention of Colonel

Squier, he directed me to prepare a memorandum of what should be done and propose the exact wording of the legislation that it would take to get us what we needed.

The subject was "Aviation Development, United States Army." In it I gave vent to my conviction that action had to be taken swiftly for the sake of American air power. I pointed out in great detail the immediate and general needs of the Aviation service of the Army and recommended reorganization of the Washington office of the Aviation Section, legislation for more efficient use of appropriations then available, a survey of all military posts to determine their use as flying schools and air stations, and legislation which would lead to the establishment of a flying corps, separate and distinct from any other corps or branches of service.

I ended my letter by saying that I hoped that it would be "considered as a thoroughly serious and sincere exposition of views and recommendations submitted solely in the interest of the service, without partiality, favor, or affection, or with any intent to discredit or criticize." General Funston endorsed my letter, and included a sentence directed at Billy Mitchell and his young coterie whose inept staff work he had seen during the time of the Mexican expedition. He said, "There can be no doubt but that the recommendations of officers of higher rank and more experience in the Army will receive more attention from all concerned than will those of young officers who, perhaps, have been advanced to an anomalous position not usually acquired by officers of their years of service."

There is no doubt that my letter and General Funston's endorsement stung the War Department in general, and Billy Mitchell in particular. Colonel Squier was not surprised, because I had told him beforehand what I was going to do. It was a direct slap at Mitchell, because, as Squier's deputy, it was his job to make the kind of recommendations I had made. However, Mitchell could not be blamed for his inexperience.

To his credit, Mitchell was sensitive about his not being able to fly, and decided to become a pilot while I had been working in his outer office on temporary duty. Since he was then too old (thirty-six) to attend our Army school at San Diego, he took flying lessons from a civilian pilot at Newport News, Virginia, on Sundays in the fall of 1916. By March, 1917, he had acquired a total of fifteen hours in the air and thirty-six landings, which, according to his instructor, were "controlled crashes" at best. The lessons cost him about $1,500 and with this newly gained expertise he fancied himself an authority on military aviation from that point on. However, his failure to anticipate and plan for the purchase of land for airfields in the best flying areas in the United States was enough to convince Squier that Mitchell lacked the necessary training and experience to develop the basic precepts, principles, and policies required to meet all of our military air war plans.

When my letter reached Washington, Billy must have known that his days were numbered insofar as his usefulness as Squier's deputy was concerned. In March Mitchell asked for orders detailing him for duty as an observer of military aviation in Europe. As soon as Mitchell left, I was ordered to Washington to take his place.

I was to see and hear more of Mitchell in the future. While serving with Pershing in Mexico, I had already formed a suspicion of his knowledge, experience, and accomplishments. During the time the 1st Aero Squadron was in Mexico, he had signed a letter for the Chief Signal Officer complaining of my "unauthorized" purchases of civilian gasoline to keep our planes in the air. General Pershing fired back a reply to the effect that airplanes, unfortunately, would not fly without gasoline and that he would authorize its purchase whenever and wherever necessary to keep Foulois' planes flying. I appreciated Pershing's backing and wondered why Mitchell, with eighteen years of service, would dispatch such a letter to an expeditionary force commander. When I arrived in Washington I checked the results of Mitchell's staff work during the months he had been assigned as Colonel Squier's deputy. The correspondence he had generated showed me that he was uncommonly good at putting words on paper. My study of his files, however, also showed that the excellence of his language was not necessarily followed by the same quality of actions. I had to begin with the basic assumption that fighting squadrons were going to be needed and had to be organized, trained, and equipped. We recommended, and the Chief of Staff authorized, the activation of twenty squadrons: five squadrons forming part of tactical organizations in the States, twelve for cooperation with United States coast defenses, and three squadrons at overseas stations.

On March 29, 1917, the first plan for the production of aircraft in the event of war was drawn up at a meeting of the National Advisory Committee for Aeronautics. Colonel Squier detailed me to act for him in the deliberations with the NACA. We came up with a program calling for more than 19,000 training and service planes for the Army and Navy. It was approved by Colonel Squier, but he couldn't sell it to the Secretaries of War and Navy. Our program had been altered in the forwarding process and now called for an appropriation of $54 million to support 16 aero squadrons, 16 balloon companies and nine aviation schools. We foresaw a force of 1,850 pilots and 300 balloonists and a training program that would turnout 650 pilots and 300 balloonists annually.

There is always a lag between military plans and the appropriations from the Congress to carry them out. When we submitted our plan on March 29, 1917, we did not know then that only eight days later the Congress would declare war. Earlier that month American opinion had been dramatically changed by the publication of an intercepted German message to the Mexican government, which proposed an alliance between Germany and Mexico if the United States entered the war. Shortly afterward, President Wilson announced that American merchant ships would be armed, and on the same day an American ship was sunk under circumstances which, for the first time, permitted no excuse whatever on grounds of error. In the next few days four more American ships were sunk by German U-boats and six Americans lost their lives. On April 2, 1917, President Wilson sent a message to the Congress announcing that "the world must be saved for democracy," and asked for a declaration of war. In spite of an impassioned plea by Sen. W. George Norris (R-Nebraska) against it, the Congress finally agreed with the President on April 6, 1917. By an overwhelming vote, the United States committed its wealth, its technology, and its manpower to "the war to end all wars."

On the day that the United States severed diplomatic relations with Germany, a total of 293 planes were reported to be on order with American firms. Only 38 had been delivered by April 6, which should have been a warning that the sudden availability of government funds does not buy an instant air force. Dr. Charles D. Wolcott, chairman of the NACA, declared in a speech that America "had hardly made a beginning" in aviation compared to the belligerents. He added that "though millions may be available for a specific purpose in time of great need, no amount of money will buy time. Even the most generous preparations do not open up the years that have passed and enable us to lay carefully the foundation of a great industry and a great aerial army through the education of engineers, manufacturers, teachers, and all the wide variety of personnel required."

These words summed up the entire American military aviation problem so well that it is a pity they were not remembered twenty years later. The United States, in its everlasting isolationist apathy, had been so badly outdistanced in aeronautics that we had not a single plane that could fire a gun or drop a bomb. There were none that could attain the speed or boast of the maneuverability of those in the air for both sides in Europe. The only thing America had to offer at that point in time was an overwhelming optimism and blind faith in Yankee ingenuity.

One of the first major steps taken after war was declared was to form a special Committee on Production, which later became the Aircraft Production Board. This board, composed of men with wide business experience, found out quickly that there was no such thing as an aircraft industry. "The industry may be divided into two parts," their report said, "the Curtiss Airplane and Motor

Corporation, and others." Undeterred, the board set to work and began their study of Army and Navy aviation needs. Military missions sent by the French and British arrived and advised what type of equipment and how much would be needed. As a result, the board called for an appropriation of $300 million for a joint Army-Navy three-year program to train 2,500 men the first year, 5,000 the next year, and to encourage the industry to produce 3,700 planes in 1918, 6,000 in 1919, and 9,000 to 10,000 in 1920.

While this was the board's estimate of American air needs, the expectations of our Allies were more ambitious. On May 24, 1917, the following cable was received from French Premier Alexander Ribot:

> It is desired that in order to cooperate with the French Aeronautics, the American Government should adopt the following program: The formation of a flying corps of 4,500 airplanes—personnel and matériel included—to be sent to the French front during the campaign of 1918. The total number of pilots, including reserve, should be 5,000 and 50,000 mechanics.
>
> Two thousand airplanes should be constructed each month as well as 4,000 engines, by the American factories. That is to say, that during the first six months of 1918, 16,500 planes (of the latest type) and 30,000 engines will have to be built.
>
> The French Government is anxious to know if the American Government accepts this proposition, which would allow the Allies to win the supremacy of the air.

This message marked the beginning of a program but it also pointed up the first of many misunderstandings that would follow. The French had been led to believe that all they had to do was ask and the planes would be quickly forthcoming from the United States. But the request served a useful purpose. It completely shattered all previous American plans and set machinery in motion for an unprecedented expansion in military aviation. The French request was too ambitious from a practical standpoint and this was highlighted by the absence of any specifications as to types of aircraft. It was no surprise to me to learn later that Billy Mitchell, our American military air observer in France, claims to have drafted the requirements in collaboration with the staff of the French Army Group of the Northeast. Marshal Joffre and his mission were then in the United States, representing the French government, and this independent action of Mitchell's confirmed in my mind that this officer was in a position to do great harm if allowed to deal at the highest level of government on air matters without control or advice from his superiors. When on duty in France in December, 1917, Maj. Townsend F. Dodd, General Pershing's Air Officer, informed me that Lieutenant Boulanger, Dodd's liaison officer in the French Air Ministry, was initially responsible for the action taken by Premier Ribot.

It was more than a month after war was declared before a leader was chosen to take over the American commitment to France. Gen. John J. Pershing was appointed Commander in Chief of the American Expeditionary Forces in May, and immediately began to select his staff. In my mind, no finer man could have been found for the job.

Once his appointment was confirmed, Pershing worked day and night on the task of building an army of half a million men to lead in France. He had seen the Ribot cable and had called Squier and me in to discuss it. As a result, I was given the job of taking the broad figures in the French request and amplifying them into a detailed program. As Pershing noted in his memoirs, the Ribot message "in its appeal for such a large number of aviation personnel and airplanes was really a most convincing confession of the plight of the Allied armies. But more than that, it strikingly brought home to us a full realization of our pitiful deficiencies, not only in aviation but in all equipment."*

I dived into my work with all the energy I possessed. I held and attended conferences day and night. I made up charts,

* John J. Pershing, *My Experiences in the World War,* Frederick Stokes Co., New York, 1931.

tables, and schedules, and tried to fit all the details of our enormous aviation requirement together. I asked myself and others hundreds of questions.

"How many student pilots should we put into a flying training program to get 5,000 out?" "Where should training fields be located?" "What length of time should it take to train a mechanic?" "How much gasoline will be needed each month?" "How much money will it cost to house, feed, and train a squadron?" "What equipment does a squadron need?" "How many men of what skills should be in a squadron?"

With no experience of any magnitude beyond a single squadron to guide me, I had to rely on my own judgment based on the trial and error of the days at San Diego, Fort Sill, Fort Sam Houston, and Mexico. There was no time for balancing and adjusting. I could not trust the estimates of those who were not fliers.

The final figures surprised and shocked everyone—including me. Beginning with the French requirement to have 4,500 American pilots in France by the spring of 1918, I figured that beginning in August, 1917, with 540 graduates, pilots were to be turned out from the primary training schools until 4,650 were trained by May 15, and 6,210 by July 15, 1918. My plan called for advanced training to begin on 2,500 De Haviland-4 planes and "high-speed training" on 800 each of the SE-5, Sopwith and Spad types. "It is entirely within our ability," I wrote, "to meet the request of France, provided immediate action is taken toward building the 500 primary training machines necessary."

Beyond the total of 4,900 planes for training, I calculated the need for 12,000 combat planes and 24,000 engines by June 30, 1918, which included 6,667 fighters, 4,000 reconnaissance, and 1,333 bombers. I could not immediately specify the exact types because I had no information from our military air observer in France, Billy Mitchell. Although he complained bitterly in later years that he had send detailed information back to the States, all I can say is that I had the task of carrying out the plan for the Ribot request in the United States, and I never received a single report or recommendation from Mitchell. However, this lack of information from Mitchell didn't worry me a bit, as I soon began to get all the information I needed through Dodd, and later from the Colonel Bolling Mission in July and August, 1917.

My job did not end with the tallying up of men, planes, equipment, and money. The next task lay in selling the General Staff and the Congress on the largest aviation program in history, which I estimated would require almost two-thirds of a billion dollars!

The shock of the old Army hands in the War College Division of the General Staff can be well imagined when I began my round of briefings. They were not mentally prepared for such an aviation program that put the air arm on a par fund-wise with the whole Army and Navy. The looks of consternation that came over the faces of my listeners was a sight to behold. None of them could conceive of the numbers of men and planes that I envisioned. They were tied to the ground forces, had never seen the airplane in a combat role, and had no conception of the uses to which an airplane could be put to assist in the ground war. I got nowhere with the War College Division, and so reported to Colonel Squier.

Squier knew he was bucking an immovable object in trying to convince the infantry, artillery, and cavalry generals that they should support such a fantastic (in their minds) aviation program. To his great credit, Squier went directly to Secretary Baker, saying that he felt sure the bill would get through the Congress if it were introduced. Baker, in his wisdom, took the unprecedented action of sending the proposal to the Congress without the approval of the General Staff. But, instead of giving the proposed appropriations bill to the House Committee on Appropriations, Baker gave it to the chairman of the House Military Affairs Committee, Julius Kahn, of California.

Squier, by now a general, and I, appeared before the Kahn committee. Using the Ribot cablegram as a basis, Squier made the following statement:

> The foregoing cablegram is based on military necessity, and from a military point of view, is conclusive and without question.
>
> The French government, without doubt, urgently needs vital assistance from the United States before the spring campaign of 1918.
>
> We can produce, in the United States, the 5,000 pilots and 50,000 mechanicians by May 15, 1918, provided the funds are forthcoming for the manufacture and purchase of training machines heretofore mentioned.
>
> We cannot produce or provide the necessary service airplanes and engines for these 5,000 pilots unless funds are also immediately forthcoming for their manufacture and purchase.
>
> The estimate for service airplanes and engines is based upon the cablegram from the Premier of France, in which he states that this country should construct 2,000 airplanes and 4,000 engines each month. Based upon this figure there should be produced by the United States 12,000 airplanes and 24,000 engines during the period January 1, 1918, to June 30, 1918.
>
> It is my opinion that the Premier of France is fully cognizant of the military needs of this country, and would not make such an important request unless he believed it of absolute military importance to the cause of the Allies.
>
> It is further my personal opinion that a country, which has been carrying the burden of the greatest war in history is fully cognizant of the enormous military value of the airplanes in actual war, and when that country asks the United States to assist in maintaining the supremacy of the air by the utilization of its enormous wealth and resources, I do not believe that any official, military or civil, in this country, is in a position to say that such assistance should be withheld, unless this country intends to act in a perfunctory manner in its cooperation with its Allies in war.
>
> I, therefore, urgently recommend that the estimates, as submitted, be approved, in order that this country may be in a position to render valuable and necessary assistance, to its Allies with the least delay.

The committee met a second time to hear the representatives of the military mission of Great Britain, France, and Italy, who supported our claims in every way. Their appearance marked the first, and so far as I know, the only time that officers of Allied armies appeared before a congressional committee.

The committee reported the bill unanimously and added one vital amendment to the effect that $640 million would be appropriated to put the program into effect. In their report the committee noted that "English, French, and Italian officers appeared before the committee in addition to our own and all concerned concurred in the conclusion that the control of the air was at this time at least the most important thing that could be done. In fact, it was boldly stated that the control of the air would naturally result in reducing to a large extent the number of men necessary in the trenches."

The bill requested an appropriation of $640 million—the largest for a single purpose in the history of the American Congress. This was an average levy of $30 for the head of every family in the United States. Drawn up in six weeks, it passed through the House Committee on Military Affairs in one week, was put through the House in a single day, through the Senate in another week, and was signed by the President on July 24, 1917—fifteen weeks after the declaration of war.

Money, the magic lubricant for the wheels of American industry, was now available; but that did not mean instant availability of the tools of war. The complete lack of experience on the part of our nation's industrialists in aviation needs, their tremendous optimism, and the absence of any organization to marshal their resources, were, in combination, to produce a national fiasco that should have taught us a valuable lesson.

It is almost impossible to realize now the utter unpreparedness of the United States in April 1917. Germany had entered the war with nearly 1,000 airplanes, France with about 300, and England with about 250. We not only had less than 100, but ours were already obsolete and not a single one was suitable for combat.

Adequate manufacturing facilities for the production of airplanes and allied equipment of a war type did not exist. Several companies and individuals had been producing very simple planes and engines of limited capacity in a small way. But this production, often carried out in garages and barns, had really not passed the experimental stage, and had not developed with any speed because there existed neither commercial demand nor adequate government support. The 59 planes which had been delivered during the eight years prior to 1917 had been supplied by four different makers, the largest number ordered from anyone of them being 22.

Before I had begun to work on the plan which ultimately led to the $640-million appropriation there was no definite understanding as to how much aircraft equipment would be required for the use of the Army or Navy and therefore no program to work to. The initial step taken was to invite aviation experts to the United States from England, France, and Italy in April 1917. Then in May we sent 100 expert American mechanics overseas to gain experience in European engine and aircraft plants. On June 17, 1917, a special technical mission, headed by Maj. (later Colonel) Raynal C. Bolling, sailed for Europe with instructions to investigate and recommend the types of planes and equipment that should be produced in the United States.

The first and biggest problem was the choice and design of the aeronautical equipment which should be adopted for manufacture in this country. Time was not available for the experimentation required to develop purely American planes. It was obvious to us that an intelligent decision regarding the types of European planes or engines which should be put into production could not well be made anywhere but in Europe. We knew that it was hopeless to attempt to weigh the arguments of the different competing representatives of foreign companies then visiting the States. The state of the flying art was progressing so fast in Europe under the impetus of war that studies of the many types of equipment would have to be made at the front by experts aided by the best counsel obtainable from each of the Allied countries.

General Squier instructed the Bolling mission to investigate both industrial and military conditions with particular emphasis on the rapid development of requirements at the front. Being separated from the scene of operations by 3,000 miles of ocean, we felt it essential that Bolling should *anticipate* developments so that by the time planes and equipment could be put into production, transported overseas, and delivered to the front, they would not be obsolete and thereby rendered militarily useless.

One of the first actions we took after the $640-million appropriation bill was passed, was to organize three divisions within the Air Service: The Equipment Division for the design and production of aeronautical matériel, including balloons; the Construction Division for the preparation and construction of training fields; the Air Division for the training of pilots, observers, and gunners. While officers were being sought to man these new offices at home, and Bolling's group was negotiating with the Allied governments in France, we arranged to have Townsend F. Dodd relieved from duty as Air Officer at General Headquarters, AEF, and given the task of coming up with plans for the actual use of our combat units on the front. Because of these reassignments overseas, a senior coordinator was called for by General Pershing's staff, and Brig. Gen. William L. Kenly (not a pilot), then on duty at San Diego, was sent to France, and assigned to duty as Chief of Air Service, AEF. His task was to obtain closer coordination between activities in the zone of advance with the combat forces and activities in the zone of communications to the rear. Mitchell, promoted to lieutenant colonel in May, 1917, was made his chief assistant in charge of Air Service activities at the front and Bolling took over the job in charge of Air Service activities in the line of communications at the rear.

In June 1917, I held the rank of major, and on July 24, 1917, was promoted to the rank of brigadier general. This meteoric rise was not owing to any superior wisdom on my part, but was done for me and a few others to prevent any gross inequities developing whereby the regular officers would be outranked by civilians getting direct appointments to high ranks. All promotions were temporary. We regulars were grateful but nonetheless a little surprised to jump two and three ranks overnight. The theory was that the regulars were going to be thrust into positions of great responsibility and needed the commensurate ranks to get our respective jobs done. The difference between the regulars and some civilian appointees who were made instant colonels or generals was that most of us weren't sure we should be promoted so far and so fast. On the other hand, the civilians, most of whom were businessmen who had kept their political skirts clean or at least the proper color, thought they should have been commissioned a grade or two higher than they were. This lack of personnel planning was just one of the many lessons of World War I that, unfortunately, was also one of those forgotten in the two decades following the "war to end all wars."

Another of the problems that soon developed was organizational. While we revamped the office of the Chief Signal Officer into the three divisions quickly, nothing could be accomplished without some joint organization to coordinate the requirements of the Army and Navy, and another to ride herd on American industry to turn out the planes and engines we needed. Therefore, in May 1917, a Joint Army and Navy Technical Board (with myself as chairman) was formed "to standardize, so far as possible, the designs and general specifications of aircraft, except Zeppelins."

Another board formed the same month was called the Aircraft Production Board and was purely advisory. It was made up of industrialists who were commissioned in various grades of captain through colonel.

While the Aircraft Production Board had no authority to commit the government, it was continuously active in the formulation of programs, and numerous contracts for planes and engines were placed on its recommendations.

The famous Ribot cable from which the whole aircraft production program was later derived was not referred to our General Staff. It went to the Joint Army and Navy Technical Board of which I was chairman, where it was approved and sent to the Secretaries of War and Navy. The Secretary of War then directed that the Chief Signal Officer be informed of America's adoption of the French suggestion.

The Aircraft Production Board then proceeded to tell the public through the newspapers of the great fleets of American planes that would soon darken the skies of Europe. This publicity campaign led the public to believe that the war would soon be won. Great harm was done by the glowing assurances appearing in the papers, and it made me cringe to see such headlines as GREATEST OF AERIAL FLEETS TO CRUSH THE TEUTONS in the *New York Herald.*

I was relying chiefly on the Bolling Aeronautical Commission for my facts, and on August 15, 1917, I got Bolling's first major report:

> First. The United States must first provide itself with all airplanes and engines required for training purposes in America.
>
> Second. The United States must next provide the airplanes and engines necessary for use strictly in connection with the operation of American forces in the field. It is best known in Washington what will be the size and composition of the American forces in the field at any given dates in the future. You have the information as to the number of types of airplanes required in direct connection with military operations of these forces. We have learned nothing to change the views on that matter which were held by Major Foulois when we left Washington.
>
> Third. After these first two considerations comes the American program of putting into the field next year air forces in excess of the tactical requirements of its army in France. It is greatly desired that the United States shall do this. Such air forces should consist of fighting airplanes and bombers.

Bolling described the conditions of European production and then advised:

> In our opinion these American needs may be divided into two periods: First period, from the present time to July 1, 1918. Second period, subsequent to July 1, 1918. With every confidence in the ultimate performance of the American production program our investigations of production

> experience over here and of the sea tonnage situation have convinced us that airplanes and engines produced in America cannot be actually delivered at the front in any great quantity prior to July 1, 1918. Subsequent to July 1, 1918, we believe that American production will not only take care of our needs, but may become a large factor in maintaining the air forces of our Allies. In considering the period between now and July 1, 1918, due weight must be given to the experience of all foreign countries and manufacturers in the delays in airplane and engine production which were not and could not be foreseen. Only at close hand can one appreciate how many and how great those delays have been....
>
> After long and careful consideration of this subject, we and all others here have come to the very strong conviction that most of the airplanes and engines for American use at the front and for our training here between now and July 1, 1918, must be produced either in France or Italy where effective and successful methods of production are already in full operation. Because we consider this imperative and absolutely essential to prevent failure of our air campaign next year, an arrangement has been made with the French Government under which they are to produce for us the airplanes and engines shown on the attached table which also includes a schedule of guaranteed deliveries....

Bolling enclosed a schedule of the approximate prices of the planes being bought and a draft of the proposed agreement between the American and French governments. General Pershing, who had arrived in France in June, signed a formal contract on August 30, 1917, for 5,000 planes and 8,500 engines. The French had entered the agreement with the express understanding that the United States would furnish the machine tools and raw materials. Needless to say, the planes and engines wouldn't get built without raw materials. The raw materials were in the forests, mines, and mills of the United States, an ocean away. Only a miracle could have enabled that contract to be carried out in a few weeks and there were no miracles to be had.

Bolling, anticipating production delays in the United States, placed orders overseas. He recommended production of British, French, and Italian planes and engines by American firms. He specifically recommended British trainers and pursuits, De Havilland reconnaissance and day bombing planes, Spad pursuits, and Caproni night bombers. LeRhone, Hispano Suiza, Gnome, and Rolls-Royce engines of various sizes and horsepower were also recommended.

It was decided that the Curtiss JN 4-D with the OX-5 engine and the Standard SJ-1 with the Hall-Scott engine would be produced as elementary training planes at home. Advanced training planes to be produced in the United States as originally selected were the Bristol Scout, the Thomas-Morse S-4, Curtiss JN4-H, and JN6-H models.

The first types of combat planes selected for production were several models of the Spad, the Martinsyde, the De Havilland-9, the Caproni, and the Handley-Page. Subsequently, the Martinsyde and the Spads were rejected, the production of the DeH-9 was first limited and then postponed; the Caproni and the Handley-Page were not treated as a part of the immediate program. The plans for home production centered on the DeH-4 and the Bristol Fighter with the 12-cylinder Liberty engine.

The Bolling recommendation for engine production centered on the Hispano-Suiza, the Gnome, and the Rolls-Royce. However, he concluded his engine recommendations by noting that the developments in connection with the Liberty engine in the United States "are probably the most important consideration in this question of engine production." In hindsight, the development of the Liberty engine became one of the bright spots in the history of World War I aviation. Called an engineering marvel by some, it was also the cause of bitter postwar accusations that its development was made at excessive cost to the taxpayers and to the great profit of the Packard Motor Car Company. Actually, the idea of the Liberty was born in 1914 when Packard decided to go into the business of manufacturing aircraft engines and then received a Mercedes engine for study that had been used by the Germans in one of their planes. Taking certain features from that and other engines, a Packard engine was designed in 1915 that developed 140 horsepower and then another that was rated at 200 horsepower.

The work of developing these engines had been under the direction of Jesse G. Vincent, then vice president of Packard in charge of engineering. In May 1917, Vincent went to Washington for the purpose of selling the idea of producing these heavier engines in large quantities by enlisting the aid of other automobile engine manufacturers. We did not believe the 200-horsepower engines were adequate for our planned needs on the Western Front, and advised Vincent to come up with a design for greater horsepower with lighter weight per horsepower. Vincent teamed up with E. J. Hall of the Hall-Scott Company, and the two of them formulated a simple, shrewd idea. Instead of a cast block containing multiple cylinders, they designed a single cylinder which could be put in different crankcases to make engines of four, six, eight, or twelve cylinders. If successful, they could thus turn out engines ranging from 135 to 410 horsepower. The great advantages to this idea would be the interchangeability of parts and the possibility that the cylinders could be mass produced in many locations.

The brilliance of the Vincent-Hall idea was in the elementary fact that our engine-power requirements would increase from about 250 to 400 horsepower or more in the time span we envisioned. They centered on the eight-cylinder version first, made extensive drawings of it, and labeled it "U.S. Standard Aircraft Engine." It was later named "Liberty" by a Navy admiral named David Taylor, one of the Navy's outstanding aviation advocates of World War I days.

To their eternal credit, Vincent and Hall managed to go from drawing board to flying line in ninety days, and the first 12-cylinder Liberty engine took to the air on August 29, 1917. It was the first engine since the radial built in 1903 by Charles

Manley for the unsuccessful Langley *Aerodrome* to weigh less than two and a half pounds per horsepower. As chairman of the Joint Technical Committee, I formally approved the Liberty engine concept based on its performance, and authorized its production.

Another of my many tasks during the spring and summer of 1917 was to select the sites for the airfields we needed for training. Based on my experience, I laid down three simple criteria: the fields had to be near a railroad, near a reasonably large population center for the acquisition of labor, and contain 640 acres (a mile square, if possible). I traveled all over the country and made decisions that have had a lasting effect not only on military aviation, but on a dozen metropolitan areas. Examples are Wright-Patterson Air Force Base near Dayton, which grew from the original McCook Field site I selected; Selfridge Field near Detroit; Call Field near Wichita Falls, Texas; and Chanute Field, Champaign, Illinois. I had previously recommended other sites which were later named Bolling Field in Washington, D.C.; Kelly Field in San Antonio, Texas; Langley Field, new Norfolk, Virginia, and Post Field at Fort Sill, Oklahoma.

At this time we were running into serious problems developed in plane production. The major issue was lumber, for the planes of those days were constructed mainly of spruce, fir, and mahogany instead of metal. And nothing better illustrates the newness of aircraft production in this country than the problem that had to be solved before the magnitude of the lumber problem was understood. The knowledge of airplane requirements had to reach the loggers, the sawmills, the cut-up plants, and follow through the processes of drying, sawing, and use of the lumber in the plane factories.

The principal difficulty in the airplane itself was to procure suitable material for the wing beams and struts. Because of the stresses developed in a plane in normal flight and especially in combat maneuvers, only the most perfect straight-grain wood was acceptable.

The typical airplane being built in those hectic days required wing beams made out of one piece of wood. This meant that suitable lumber had to be extra long, thick, and perfect. The only lumber accepted in this country for wing production was spruce from the Pacific Northwest. And enough had to be found to provide for the entire air program of the Army, as well as the Navy and our Allies.

A typical airplane used about 500 feet of lumber. This meant that it usually took about 1,000 feet of rough lumber to get the required footage that had no cross grain or spiral grain. We soon found, however, that we had to obtain as high as 5,000 feet per airframe because of the imperfections in the lumber, difficulties in getting proper inspection at the mills, and faulty handling in transit and in the airplane factories.

We soon learned that we had to find alternate courses of action. We began to substitute fir for spruce, and we authorized the construction of wing beams from small pieces of lumber by splicing and laminating. I remembered the difficulties we had had at Columbus, New Mexico, in 1916 with our laminated propellers and didn't like the idea, but I soon learned that I was going to have to accept compromise with quality if we were ever to get any American planes into the air over Europe.

Another compromise developed over the fabric used on planes. When we entered the war, linen was used exclusively and there were only three sources of flax: Belgium, Russia, and Ireland. The Belgian supply was cut off immediately; Russian flax was always difficult to obtain. This left Ireland as the only source of flax for airplane linen.

Since every plane used from 250 yards to 500 yards of fabric, an impossible shortage soon developed. We turned to cotton fabric. This caused additional problems because the dope used on linen wouldn't work as well on cotton. New dope had to be developed. However, before the war was over, we got satisfactory grades of cotton cloth and nitrate and acetate dopes into production.

The requirement for aircraft guns presented more

production difficulties. Although a machine gun had been fired from an airplane in this country in 1912, nothing further was done until after war had actually started. The French had a few heavy airplanes equipped to carry machine guns in 1914, but it was almost a year later before machine guns were installed on planes for combat use. The pilots and observers overseas, however, were armed with various weapons such as service rifles, automatic rifles, automatic pistols, and shotguns. They were also equipped with grenades and darts, which were intended to be dropped on their adversaries.

The invention of a synchronizing device by the Germans which would allow the pilot to fire through the propeller changed all this, and considerably altered air warfare. This fact, more than any other, took the airplane out of the defensive role of an observation vehicle and put it in an offensive role where it could fight back.

Aircraft bombs were another item where we were caught short. We authorized the Bureau of Ordnance to experiment and produce suitable bombs, release mechanisms, and bomb sights. Again it was a case of being too late in starting. At the beginning of our entry into the war we had nothing. When the war ended, we were just getting the hang of it. However, bombs of various weights from 50 to 1,000 pounds were produced. The 100- and 250-pound sizes were used mostly. Fragmentation and incendiary bombs were also being produced in quantity at the end of the war.

There were other items that had to be developed to keep pace with the Germans. These included pyrotechnics (flares) used for signaling, aerial photographic equipment, aviators' clothing, oxygen, radios, and parachutes. Although all of these things had been recommended by me and other pilots in the years preceding the war, nothing had been done to develop them. It took a war to bring the logic and the urgency of these inventions to the attention of the General Staff, the Congress, and the American people. By October of 1917 I felt that I had done all I could to get the wheels of American industry rolling.

When Pershing had gone to Washington early in 1917 and was forming his staff, he had called me into his office along with Colonel Squier. "Foulois, you and I have served together before and I've always liked the way you try to get the job done, come hell or high water, on the ground or in the air," he said, giving me one of those rare smiles. "What are your plans for the war?" he asked.

1 looked at Squier and he took the cue. "General, if you're thinking of taking Foulois with you, I'd like to register a dissent before you make your decision."

The smile faded from Pershing's face. He stared at Squier like a schoolteacher who had just been interrupted during a lecture. Squier rushed to explain.

"General, I know you need staff officers," he countered, "but I need Foulois here to get our air program going. I called him in to replace Mitchell and he has put our air plans together practically single-handed. I'd like him to stay here at least six months and then he can be made available."

Pershing's brow wrinkled. He looked at me. This time I took the cue.

"Yes, General, I think I should stay here," I told him. "If you want me in six months, I'll be honored to be on your staff in France."

Pershing nodded. He asked me for a nomination and I immediately recommended Townsend F. Dodd, then only a major, whom Pershing had known as my deputy in Mexico. Pershing agreed to take Dodd but with the full understanding that I would join him in the fall. He bade us good-bye and we saluted and turned to go. "See you in France, Foulois," he called after me as I shut the door.

"Yes, sir," I answered. It was an order as far as I was concerned. I had not realized how much time had passed when Squier called me into his office in September. "Benny, I think it's time for you to go to Europe," he announced.

I nodded. "Does General Pershing still want me?" I asked. "He not only wants you, he needs you, Benny," Squier replied.

"He's having trouble with Mitchell who is constantly fighting with Dodd. It's too bad that Mitchell outranks Dodd, because Dodd is the better man. Since you're a general now and outrank Mitchell, you don't have to put up with his foolishness. I want you to turn over your office to your deputy and leave as soon as you can."

I knew what Squier meant about the Mitchell-Dodd affair. Dodd was a quietly efficient officer who knew as well as I did what was needed to get an air force into the air. Mitchell, a compulsive and persuasive talker, only *thought* he knew. Being loud and a rank higher, Mitchell held a winning hand. If he had been a doer instead of a talker, we might have had a semblance of an air organization in Europe in the fall of 1917. To this day I don't know what useful contributions he made to our air effort in the first six months he was in France. The only thing I did know for certain was that he made trouble for everyone I sent over to help him.

To give a clue to the personality of the man who had placed himself in a position of grave responsibility but who did not have the proper background for it, I refer the reader to Mitchell's memoirs.* In them he quotes Dodd as saying that the Chief Signal Officer's office was a "madhouse" and that "the Army… was perfectly helpless to cope with the aviation problem."

Dodd, an old friend, denied these statements. Instead, he told me later, Mitchell had gotten things mixed up. "I told him that things in *France* were so fouled up," Dodd said, "that it looked as if German saboteurs had been at work." Mitchell took an immediate dislike to Dodd, and the feeling was mutual. Thereafter, Dodd volunteered to work with the Bolling Commission in London, and Mitchell happily approved the transfer. I shall always believe the loss of Dodd owing to this personality conflict set us back weeks in getting things organized in France. When I sailed for Europe in October, I

* *Memoirs of World War I* by Brig. Gen. William Mitchell, Random House, New York, 1960.

knew that one of the biggest problems I would have confronting me was a single American: Billy Mitchell.

As soon as it was decided that I would serve on Pershing's staff, the question arose as to my title. Squier proposed to the Secretary of War that I relieve Brig. Gen. William L. Kenly, who had been sent over in August as Chief of Aviation of the AEF. It had been agreed with General Pershing before he left that the Air Service would have to be separated from the Signal Corps in France, and organized and maintained as a distinct force. Kenly's job was, in Pershing's words, "to coordinate our training with that at home." When Kenly arrived, he immediately had a run-in with Mitchell and got rid of him by putting him in charge of aviation in the zone of advance. Bolling was placed in charge of the zone of the interior and served as Pershing's representative on the newly established Inter-Allied Aircraft Board.

Kenly was a good staff man, and tried to bring order out of the organizational chaos that he found, but not being an aviator had its disadvantages. He could not put up a knowledgeable defense against Mitchell's arguments and demands. Mitchell *sounded* like an expert, and, I suspect, Kenly gave in to him. In any event, I was ordered to replace Kenly, who, in turn, was assigned initially to staff duty at General Headquarters, AEF, and in May 1918, returned to the United States for duty as director of Air Service Operations.

My orders to Europe authorized me to choose my staff. When I walked up the gangplank in late October, I had a staff of 112 officers and 300 enlisted men. There wasn't a single man among them who did not have an outstanding record of efficiency and accomplishment behind him. No officer ever sailed off to war with a finer team. Little did I realize that my team would have to fight Billy Mitchell, the Army General Staff, the United States Navy, the British, the French, and the Italians before we would ever get to fight the Germans.

Representing U.S. airpower, officers of the 1st Aero Squadron, San Antonio, Texas 1915. (Left to Right) Lt. T.S. Bowen, Lt. J.E. Carberry, Lt. C.G. Chapman, Capt. B.D. Foulois, Lt. T.D. Milling, Lt. I.A. Rader. U.S. Air Force photo.

Gen. Benjamin D. Foulois (center) with French pilot (left) who flew him to different squadrons during inspections in their French Caudron C-23 1917. During World War I, most U.S. Army fliers flew aircraft purchased from France. Photo courtesy History of Aviation Collection, University of Texas – Dallas.

Chapter Ten

The War Behind the War in the Air

I arrived in France on November 12, 1917. If anyone had told me that day that the war would all be over a year later, I would have been sure to ask, "Who won?" because the American Army in France certainly didn't look like a victorious army from what I saw. Although I had expected some lack of organization, I was not mentally prepared for complete chaos when it came to the Air Service.

General Pershing, as soon as he had arrived in June, had quickly realized that his success as a commander would be impossible without efficient staff work. After studying the British and French staffs, he adopted an organization largely patterned after that of the French. His staff had three main divisions: general, technical, and administrative. The general staff was divided into sections which varied depending upon the size of the command.

To train staff officers Pershing set up a staff college near Chaumont and directed that only graduates of this school would be assigned to fill staff vacancies. The result was that Pershing's staff quickly became an efficient team in every sense. To my regret, however, the Air Service could not boast of being members of his team. Neither Bolling, Dodd, nor Mitchell attended the staff school. Bolling was doing his business by mail and cablegrams with the State Department in Washington, Dodd was trying to persuade Pershing's staff to give him the supplies and facilities he needed to get the training of flying cadets under way, and Mitchell, then a General Staff military attaché, was running all over France raising the devil with everyone he met. Mitchell spent most of his time talking about forming huge tactical and strategical aviation units, the latter intended to fly deep into enemy territory to attack airdromes and factories. There was nothing wrong with the idea, and I certainly agreed with his concept since it was just what I had envisioned in 1907 when I wrote my thesis at Fort Leavenworth. The one fundamental that Mitchell did not seem to understand, however, was that it took time to get the planes manufactured, the men trained, and a supply and maintenance system organized to launch any kind of air operations. It was my job to get the men, planes, and system working so that we could carry out such a grand scheme.

I took over as Chief of Air Service, AEF, officially on November 27, 1917. After touring all over France to see what the true situation was, I returned to Paris and announced the organization of my headquarters. I relieved Bolling as assistant chief of Air Service and put him on duties which permitted him to give his full attention to perfecting the arrangements he had initiated as head of his commission. I decided to keep my headquarters in Paris, and allowed Mitchell to remain at Chaumont as Air Commander, Zone of Advance. However, I was already dubious about letting him have any further assignments with the Air Service because of his obvious ignorance of aviation matters and his continual harassment of my staff from the moment we set foot on French soil.

It is interesting to note what was going on in Mitchell's mind as evidenced by his memoirs published after his death.* He titled a chapter "Our Air Force Disorganized" and then stated:

> ... Just when things had begun to work smoothly, a shipment of aviation officers arrived under Brigadier General Benjamin Foulois, over one hundred in number, almost none of whom had ever seen an airplane. A more incompetent lot of air warriors had never arrived in the zone of active military

* Brig. Gen. William Mitchell, *Memoirs of World War* 1, Random House, New York, 1960.

> operations since the war began. Foulois, I am told, had orders from the President to General Pershing to put him in charge of aviation in Europe, even though he was no longer an active pilot. They say he announced before leaving the United States that he would command not only the American services but in a short time that of all the Allies as well.
>
> ... The orders to General Pershing left him no discretion, so he had to relieve General Kenly and put General Foulois in charge. While Foulois meant well and had had some experience in aviation in the United States, he was not at all conversant with conditions in Europe. As rapidly as possible, the competent men, who had learned their duties in the face of the enemy, were displaced and their positions taken by these carpetbaggers.

As far as I am concerned, this extract is proof of Mitchell's disregard for facts when it came to aviation during World War I. He said I was no longer an active pilot when I reached France, but I flew as a pilot and observer throughout my eighteen months overseas, in spite of the fact that 90 per cent of my job was on the ground. If he had bothered to check when writing his memoirs, he would have known that I flew throughout the remainder of my service until I retired in December, 1935.

Further proof of his ignorance in complaining about my "incompetent lot of air warriors" is his inference that we needed pilots at that moment when, in fact, we had no planes to fly, no organization to train them, and no facilities to sustain air operations. First priority had to be given to non-flying technical projects, and the men I selected certainly knew how to plan, organize, and construct what we needed.

For the record, I also want to dispute Mitchell's statement that "the competent men, who had learned their duties in the face of the enemy, were displaced and their positions taken by these carpetbaggers." The only officer who was displaced was Mitchell himself, who was replaced by Col. Robert O. Van Horn, a veteran field commander of the Spanish-American War and the Philippine insurrection. A thorough professional with a no-nonsense approach to any task assigned him, I sent Van Horn to replace Mitchell as Air Officer in the Zone of Advance and assigned Mitchell to command the Air Service components of the 1st Army Corps. This was done for two reasons: first, to put him under the disciplinary control of the 1st Corps commander, Gen. Hunter Liggitt and his Chief of Staff, Col. Malin Craig; second, to give him a corps air tactical unit, which had plenty of problems to solve and thus provide him with an opportunity to supplement his General Staff training with experience in practical air logistics. How this reasoning worked out will be seen later.

While Mitchell had every right to have an opinion about me and my staff, his attitude toward us made our jobs doubly difficult. The seeds of insubordination had already been sown when I relieved him, and his memoirs prove how distorted an opinion he had of himself as an expert on air matters. Here was a man who had not had one minute's flying time as a rated Army pilot since he had not yet been through any official instruction course. Information furnished me upon my arrival in France by my old 1st Aero Squadron pilots indicated that his flying in France was limited to being a passenger with French pilots who did not trust him enough to let him handle the controls. And here was a man who knew so little about the logistics of military aviation that he had not yet provided a personnel depot to receive the increasing influx of Air Service personnel, had not yet made any master list of the material he needed to operate the squadrons he was supposed to be forming, and had made no enquiries into a system of resupply once the squadrons got into combat. He had been in France eight months by the time I got there, but I could find no evidence of solid accomplishment that I could attribute to his efforts. To say I was disappointed would be much too kind a statement. I was furious at his gross incompetence and reported my fury to General Pershing.

Pershing agreed with me completely, and then said, “That's why I brought you here, Benny. I get a lot of fine words from

Mitchell when I ask for a progress report, but I never see any results. You report directly to me whenever you want to and raise hell whenever needed in my name. I'll back you even if you're wrong because I know you won't be wrong very often."

I appreciated General Pershing's encouragement, and it came at the right time because the more I saw of the Mitchell brand of planning, the madder I got. I was appalled at the contractual arrangements that had been made with the French, and at the way American money had apparently been poured into a bottomless pit. I put Dodd in charge of the Supply Section in my headquarters and directed him to make a thorough study of the liabilities and outstanding obligations of the Air Service. Within five weeks Dodd and a Major Jennings put a system into effect that enabled us to account for everything procured and all monies disbursed to pay our creditors. Jennings set up a Disbursing and Legal Section which functioned successfully throughout the entire period of operations.*

The storage of supplies, once acquired, soon presented a problem which had to be solved quickly. Since we had to buy the major portion of our supplies in and around Paris, a large warehouse was obtained at Clichy in which supplies were concentrated and then distributed to units and stations as they came into being. I also made arrangements to use the facilities of the French Air Force at Le Bourget Field for the assembly, test, and acceptance of planes and engines before sending them to units at the front or to training schools.

We used the facilities at Le Bourget until March, 1918, when German night air bombardment interfered. The German attacks were excellent examples of the strategical operations I had envisioned for the future when I had written my thesis at Leavenworth. Although we don't give them proper credit for it,

* As a result of this initial care in establishing a system of supply, accounting, and disbursing, the Air Service, AEF, was the first department to obtain a clearance from the Liquidation Commission after the signing of the Armistice.

their bombing operations against Paris, and especially against the aircraft factories and fields around the city, were extremely effective. The establishment at Le Bourget was severely damaged and caused a considerable loss of planes, engines, and supplies stored there. As a direct result of these enemy air operations, coupled with the imminent threat to Paris by German ground forces, the French immediately took action to reduce the quantities of aircraft supplies stored in and around Paris and asked us to find another site to test and accept the material received from their government. I immediately chose a site at Orly and designated it "Aircraft Acceptance Park No. 2." Thus, establishments at Clichy, Le Bourget, and Orly were the principal bases around Paris to take care of the French sources of supply.

Next on my priority list was the establishment of bases to receive and store supplies coming from the States. All planes coming from the States were to be in a partially knocked-down condition. When they were unloaded at French ports, the question arose as to whether or not we should establish assembly plants at each port where the planes could be put together, tested, and then ferried to the front or to training schools. I decided against this because of the great number of plants, equal to the number of French ports, that would be required. Instead, I decided on one huge assembly plant, conveniently located near a railroad center which, in turn, would be linked to all the base ports. The most logical place meeting all my requirements was Romorantin, located about 100 miles south of Paris and about 50 miles east of Tours. Actual construction started on January 16, 1918, and was the largest enterprise of its kind in the world at that time. It covered an area of four and one half square miles and consisted of living quarters, shops, warehouses, administrative buildings, and maintenance areas, plus an extra-large flying field for added safety during test flights.

By the cessation of hostilities, the center was 85 per cent complete and we had approximately 10,500 people working there.

Had the war continued, we had expansion plans which called for a minimum personnel strength of 47,500 and a maximum of 66,000. It is a tribute to my "carpetbaggers" that such an installation could be planned and operated in view of the unending difficulties involved in the securing of an adequate supply of people and construction materials.

The location and construction of air depots and airfields in the Zone of Advance were much more complicated and uncertain problems than that of planning similar installations in the rear. Although it was generally understood that the American Expeditionary Forces would occupy a sector of the front with general limits somewhere between Verdun and the Swiss border, the prospects of a big German offensive in the spring of 1918 gave us second thoughts. It was difficult to plan any depots and fields with any reasonable assurance that they would ever be completed and put into use by our own Air Service. But we had to take a chance and hope that we weren't going to build them for the Germans to use.

Another difficulty in planning for the advanced bases was the fact that the sector assigned to the American forces was one of the worst that could possibly have been assigned from the airman's viewpoint. The area was either mountainous, wooded, or rolling countryside with very few smooth and level areas suitable for flying fields. Those areas that were of sufficient size were located in river bottoms which were usually flooded most of the year. In spite of this, I sent survey teams out during December 1917 to investigate all possible sites in the rear of the sector to be assigned our forces to a depth of 40 to 80 kilometers behind the front-line trenches.

Since there were some completed or partially completed fields within our tentative area that the French planned to use, I held several conferences with French air officials and we agreed that if we couldn't get our own bases built in time, we could use theirs. At the same time I worked out arrangements for the establishment of air depots in the Zone of Advance. The first air depot was established at Colombey-Les-Belles. Other sites were planned but no definite steps were taken to construct any more because of the German push that Pershing's intelligence staff predicted was coming in the spring of 1918.

While my staff was working day and night on these problems, I was off finding others that we didn't know we had. From the date of the initial contract with the French that General Pershing had signed on August 30, 1917 (which provided for the French to furnish us with 5,000 planes and 8,500 engines) to December 1, only a trickle of raw materials had arrived in France from the United States. The French Minister of Munitions formally notified me that they would be unable to meet the contract because we had not kept up our end of the bargain. I knew it was a bad bargain to begin with and I guess I wasn't surprised.

Since General Pershing had personally signed the contract for the United States, I reported the facts to him and recommended we cancel it and write a new one. He agreed, and I began negotiations with my French counterparts whereby they would furnish our units with all aviation material for all large established tactical units (Army corps and divisions) as rapidly as we could place them on the French front.

It took four months of argument to come to any mutually satisfactory agreement whereby we could have a reasonable assurance that we could get a sufficient number of planes for our planned air units scheduled for activation in the summer of 1918.

I sent countless cables in February, March, and April pleading for the raw materials the French needed to get us some air power, even if American production had to be curtailed to do it. My persistence finally paid off, and by May 1 about 90 per cent of the raw materials and tools required under the old contract had arrived. The French signed the new agreement on May 3. This, I believe, was the turning point in our eight-month struggle to ensure that there would be a sufficient quantity of planes and engines to equip all the squadrons we could organize until American-built planes and engines arrived.

By Christmas Day, 1917, approximately 7,500 Air Service personnel were in France—enough for 45 squadrons. I promptly organized these men into Service and Supply Squadrons to meet tactical needs and for duty at schools and depots.

As it turned out, shipping space was the major problem and Pershing was advised that the monthly troop capacity on the ships available would be 92,000 troops—barely enough to transport the two divisions of ground combat troops plus their support elements which Pershing needed. It was a tough blow, and Pershing had no choice but to reduce the aviation personnel program in view of the probable German offensive.

There was good reason to fear a German onslaught. By the end of 1917 German hopes of winning the war through unrestricted submarine warfare were fading. But there was still time for the Germans to strike a decisive blow in the west before the full strength of American arms could be brought to bear. Many troops were available as a result of the collapse of Russia, and Germany decided to resume the offensive. Field Marshal Erich von Ludendorff, now virtually in command of the German Army, selected the Somme area for the attack, hoping that a breakthrough there would drive a wedge between the French and British and that a British defeat would cause the French to lose heart and agree to peace.

On March 21, 1918, the giant German sword fell with shattering force on the British. Their lines were pierced, and for a time it looked as though the German advance might develop into a race to the sea. By March 26, Amiens, a strategically located rail center, was in serious danger of falling into enemy hands. If it fell, continued cooperation between the British and French would be difficult if not impossible. Ludendorff hurled 20 of his best divisions into the battle, but they were repulsed and were unable to take advantage of the gap they had created between the British and French lines. Allied reserves succeeded in holding their lines, and by April 5 the offensive there was over. However, the Germans had won a tactical victory. They had advanced 40 miles in eight days, took 70,000 prisoners and 1,100 guns and inflicted 200,000 casualties. However, they also paid a high price in human life for their daring.

Despite the indecisive battle on the Somme, Ludendorff clung to his plan to destroy the British Army before taking on the French and Americans. He struck in the Lys area in Flanders on April 9; but, although gains were made, the British held. This offensive was followed by another in May to cross the Aisne River. By May 29 the Germans had reached Chateau-Thierry, less than .40 miles from Paris, and the situation looked black for the Allies. The French called on Pershing for help and he promptly responded with two divisions as the enemy was about to cross the Marne River. In a three-day battle every German effort to establish a bridgehead was repulsed, and the advance toward Paris was checked. This first major contact with American troops had changed German opinion about their combat effectiveness from one of open skepticism to healthy respect.

The same could not be said of American air units, unfortunately.

In January 1918, I moved my headquarters to Chaumont where I dislodged Mitchell from his splendid quarters and sent him to the 1st Army Corps at Neufchateau. This change of location is noted in Mitchell's memoirs with the remark that "it was bad enough having this crowd down in Paris but to bring them up near the line was worse."*

This statement was indicative of the complete disdain Mitchell had for me and the members of my staff and further shows the utter lack of understanding of what it took to get planes and pilots to him so he could carry out his grand air strategy. During the months of January and February, 1918, I had considerable differences of opinion with him on his responsibilities and authority in connection with the location of airdromes and depots and the supply system to be used.

Mitchell wouldn't agree with Col. Van Horn, who had replaced him, on the color of white paper, and I had to step in during many conferences to keep Van Horn, a usually quiet man, from tearing into Mitchell. After taking all of this I could, I knew that Mitchell

* Mitchell *op, cit.*, p, 177.

could not handle a combat command if he could not work with his superiors and the people who were trying their best to help him. My concern reached a peak while on an inspection trip near Paris in March. There the First Secretary to the American Embassy sought me out to tell me that Embassy security agents had heard Mitchell make derogatory remarks about me and my staff in cafes, bars, and other public places in Paris. This information confirmed reports that I had been receiving from other sources. I immediately composed a letter to Mitchell, of which I shall repeat excerpts here:

> My Dear Mitchell:
>
> It has recently come to my attention that you have made several very drastic criticisms of the present Air Service organization, AEF, which are directed chiefly against myself and a few of the senior nonflying officers of the Air service, and I regret very much that I have learned this from other sources than directly from you.
>
> The Commander in Chief has carefully instructed me to make every effort to coordinate and build up the Air Service into an effective and harmonious organization. Since these criticisms have been traced directly to you, you must be out of harmony with the Air Service policy, or else your criticism must be based on a misunderstanding. In order, therefore, that there can be no continuance of your misunderstanding (for I am loath to assume that it is anything worse), I shall outline in writing what, in my opinion, is the proper policy in reference to the relations which shall exist between the nonflying and flying officers on duty with the Air Service—a policy which I have, on previous occasions, outlined to you verbally.
>
> It has been approved policy of the War Department, since war was declared, that all trained flying officers, if physically and mentally qualified, should command tactical and strategical flying units, and that nonflying officers should normally perform executive and administrative duties, which do not involve flying.... Acting on that policy, I have consistently informed all trained flying officers that, as soon as they could be relieved of strictly administrative duties, they must keep up their flying in order to train up to the command of flying units at the front.

At the outbreak of the war, as you well know, there were not over a dozen trained flying officers with sufficient administrative and executive experience to be entrusted with the task of expanding a force of a few hundred officers and men into a war organization which will (soon) consist of 10,000 officers and 100,000 men, or a force greater in strength than the entire regular Army of four years ago. You must fully realize that in order to carry out such a task successfully, the Air Service would necessarily have to call upon the regular Army for a reasonable number of trained officers, not experienced in flying, but experienced in organization and the art of war. The necessity for this action was fully apparent to the War Department and when I was informed that I would be sent to France, I immediately took steps to secure a small group of nonflying regular officers of unquestioned executive ability who were excellent field soldiers, honor graduates of the Army schools, who knew each other and knew me…. These officers, when they entered the Air Service, were fully informed that the trained, technical flying officer who demonstrated his ability to administer and operate tactical air commands would receive first consideration as regards promotion or important assignments. These officers further understand that if they aspire to the command of large tactical air units, they must do a sufficient amount of flying and acquire a sufficient knowledge of the technical side of the Air Service to qualify themselves for promotion. With this knowledge in mind, these officers have served loyally and faithfully during the past four months, day after day, and without regard to hours, with a common object in view—that of building up an efficient and harmonious organization for service with, and as a part of, the American Expeditionary Force in its operations against the common enemy.

I have had the impression, and the officers herein mentioned have had the same impression, that there is a strong feeling of resentment against them, based chiefly on the theory that because they are not trained fliers they cannot organize and build up an efficient Air Service. In view of the very limited number of experienced executive officers amongst the trained flying officers, such resentment is unwarranted, unjust, is a detriment to the morale of the Air Service, and is liable to create a condition opposed to the policy of the Commander in Chief.

If your motive in making these criticisms against the organization and administration of the Air Service was actuated by a desire to improve its efficiency, then I must frankly state that your method of going about it is not in accordance with my idea of military procedure.... You, with twenty years of service in the regular Army, know full well that constructive criticism is the only kind of criticism which can build up the Air Service or any other service. You also know full well that it is your military duty to the Service, to the Commander in Chief, and to me, to immediately and officially bring to my attention anything which, in your judgment, should be corrected. You further know, with your experience in the military service, that adverse criticism of your military superiors, or the manner in which they perform their duties, expressed in the presence of other officers or enlisted men, is prejudicial to good order and military discipline, is detrimental to morale, and is an offense triable by court-martial.

If your motive in adversely criticizing my administration of the Air Service is solely directed against me personally, I am exceedingly sorry that such a feeling exists. I asked to be sent to France for active service at the front which, in any opinion, should be the ambition of every soldier. I did not, however, ask for the job of chief of Air Service, but having been assigned as such by the Commander in Chief, it should be fully understood by yourself, and all other officers in the Air Service, that so long as the Commander in Chief retains me as chief of Air Service, and holds me responsible for the efficient organization and administration, I expect loyal and efficient support from each and every officer in the Air Service.

Now, my dear Mitchell, having expressed my opinions regarding a subject which has been extremely distasteful, I sincerely hope you will believe me when I state that I desire your fullest and most friendly cooperation in the endeavor to build up the Air Service. After all, we both have but one object in view, and that is to lick the Germans in the air, and

> it is my earnest hope that we all realize that there is only one way to do it, and that is by pulling together as a team....

I imagine the letter was a shock to Mitchell's ego. It would have been to me if I had received such a letter from a superior officer. It took two weeks for Mitchell to reply and when he did he began by praising me for the spirit which had actuated my letter to him saying that "I consider it a very manly letter." He admitted that he had criticized the organization of the Air Service but said that his criticisms were "the result of considerable experience in this war and seeing the manner in which the lessons learned in it have been disregarded."

He launched into a rationale that "non-flying officers should not be entrusted with work which they cannot possibly know anything or very little about [because] it puts the lives of all in the air in jeopardy and creates an extremely bad morale among the flying personnel who had to do the fighting." He complained that his recommendations were never followed, his rank had been "juggled" to his disadvantage, he had been displaced "and assigned to a place with no command," and never received communications from me or my staff while his subordinates did. His egotism really showed through when he said that "after all the efforts I have made to further aviation in this war, which are a matter of record, to have this treatment extended, not only to myself, but to other officers, appeared entirely unjust, unwarranted, and designed to create a very bad state of affairs in aviation."

Mitchell's letter ended on a polite note expressing his appreciation for my letter "and the motives that actuated it", but I could see that I had not convinced him that he should work with me and not against me.

When I analyzed the differences between us, they invariably hinged on the relative responsibilities and duties of the flying and non-flying officers—a theme that was to stick with Mitchell until the day he died. What is so ironic about his resentment is that he was technically a non-flier himself until he was rated a Junior Military Aviator in September 1917, a rating which appeared to be in violation of Section 3 of the Act of Congress which created the Aviation Section of the Signal Corps and limited the rating to officers "with rank below that of captain." This act was still in effect in September 1917, when Mitchell mysteriously received his JMA rating as a lieutenant colonel. How he managed to do this without taking any Army instruction or passing any official flying proficiency tests still mystifies me to this day. However, once he acquired his wings by whatever means, he identified himself forever after on the side of men who flew and considered all non-flying humans from that time on as inferior beings.

Even though I had reorganized the Air Service shortly after my arrival, I felt it necessary to do so again as the American buildup gained momentum in the spring of 1918. I appointed a board of officers at Tours to study how the growing load then being carried by the Supply Section could be more equitably distributed and also speed up the development of the radio and photographic sections. The board submitted a plan for reorganization which I approved. I had asked Mitchell to work with the board and give the members the benefit of his experience in the Zone of Advance. He met with them one time and refused to return. He criticized what they were doing, but rather than discuss the supply system, sought to modify the existing policies regarding my jurisdiction over his activities in the Zone of Advance.

By the middle of May I could see that my difficulties with logistics were minor compared to the irreparable damage to our efforts Mitchell was doing with his criticism of me and my staff. I was faced with a difficult decision. Should I let his continued harassment go on behind my back or should I relieve him and ask that he be sent home? Since Mitchell had no fear of going over my head to Pershing and frequently did, I began to suspect that his great gift of gab and ability to rationalize his lack of accomplishment on the practical side of aviation were getting through to Pershing. I asked General Pershing to relieve me as chief of Air Service and put me in charge of aviation in the 1st Army where I could put my tactical experience to work and get the coming air squadrons whipped into shape.

I suggested that Maj. Gen. Mason M. Patrick, a West Point classmate of Pershing's and my former commanding officer in Cuba, would be the best man to replace me. He had commanded the 1st Engineers and was then director of Construction and Forestry on Pershing's staff. Since construction was now our immediate major problem, there was no better man in France to manage the program that we planned. My value, as I analyzed the situation, lay in my practical experience with planes and pilots—not as a manager of construction projects.

Pershing agreed with my reasoning and orders were issued on May 29, 1918, giving me Mitchell's job and making Mitchell chief of Air Service, 1st Corps. The same order designated the assignment of the 27th, 147th, and 96th Aero Squadrons to the 1st Army so he would have some flying units to command.

I knew the order moving Mitchell down a notch would make him furious and I braced myself for the inevitable confrontation that would take place when I took over his office at Toul. On June 3, accompanied by Frank Lahm, then a lieutenant colonel, I went directly to Mitchell's office carrying the orders making the changes. When I entered his luxurious (for those days) office, he greeted me coldly, like a school principal being visited by the head of the PTA.

"There's no use beating around the bush, Billy," I told him. "I'm here to take over your office, your files, and your job. You are relieved as of this moment."

The expression on Mitchell's face was pathetic. He turned gray and his jaw sagged open in shock as if I had kicked him in the groin. There was a long silence, and then he recovered. He demanded to know what his status was to be, what tactical command he was to have, and where he would be stationed. I told him that he was to be chief of Air Service, 1st Corps, and his duty station would be as I directed. He began a loud monologue about how I had been out to get him ever since I had been promoted to brigadier general. When I tried to interrupt and refute his allegations, he babbled on almost incoherently and burst into tears like an immature child. Lahm and I were embarrassed for him, but his actions only proved to me that he was indeed unfit to command.

When Mitchell recovered his composure he must have realized that it would do no good to argue with me. I was his senior now and I had irrefutable bona-fide orders from Pershing to relieve him. I asked him to stay a few days to acquaint me with his files and his plans.

"Not on your life, *General,"* he said, emphasizing the word "general" unnecessarily. "You couldn't possibly acquire the knowledge to run this office in a few days and I'll be damned if I'm going to make it easy for you."

I overlooked this obvious insubordination because I understood his bitterness. Whenever a military man is relieved of his job because, in the opinion of his superiors, he cannot handle it, his ego is badly damaged. Before I had a chance to ask him, Mitchell burst into another monologue and informed me that his entire office force, the furniture, the maps on the wall, and even the telephone belonged to him, and he was going to take them with him.

I made no comment, hoping that his training as an officer would overcome his bad judgment in talking to me like that. I asked him about the routine of his office, the chief difficulties to be overcome, and his needs regarding people and matériel. He completely ignored my questions and launched into still another tirade about the commander of the 1st Air Depot, the lack of attention given to his many recommendations, and the quality of the officers on my staff.

While Mitchell was having his catharsis, I was convinced that he was in the first stages of a nervous breakdown. No regular officer with twenty years' service would have said what he did to his superior if he were in his right mind. Frank Lahm was my witness and cast knowing glances at me as Mitchell raved on almost oblivious to our presence.

That afternoon, after a strained luncheon, I inspected the 1st Pursuit Group on the Toul airdrome. We were accompanied by a number of junior officers and, to my embarrassment, Mitchell seized on every opportunity to criticize the supply people for failing to provide this or that article of equipment. This performance was the last straw. We returned to Mitchell's office and I told him to remain in charge of his office until I returned the next day. That night Lahm and I went to see the Chief of Staff of the 1st Corps, Gen. Malin Craig. I told Craig everything that had happened and found that he already knew of my encounter with Mitchell and had ordered Mitchell to transfer his people and equipment to me without further ado. He had sent one of his staff officers, Lieutenant Colonel Heintzelman, to see that Mitchell carried out his orders.

I returned to Toul the next day and found Heintzelman arguing loudly with Mitchell. Mitchell insisted that he needed certain individuals on his staff in spite of what Craig had said. He demanded the maps and charts on the walls and again was refused. Mitchell then demanded that the desk be transferred with him because he had had it a long time and it was filled with his private papers which would take some time to remove. He offered me his adjutant's desk until I could get one through supply channels. In order to terminate any further discussion, I told him to take it since it seemed to be an obsession with him that he had to win at least one point.

After Mitchell left with his desk, I decided that I had to report all that had transpired to General Pershing. I wrote a long memo detailing the various episodes and noted that Mitchell's conduct was, in my opinion, “extremely childish; and entirely unbecoming an officer of his age, rank, and experience.” I then inserted the following paragraph:

> Colonel Mitchell, during the past year, has had considerable hard field service in France, and from my personal knowledge, has on numerous occasions performed excellent service, particularly during the past month. His actions during the past two days, however, firmly convince me that he is either mentally unfitted for

> further field service or is incapable of working in harmony with me in carrying out the policy of the Commander in Chief.

In my final paragraph I recommended that Mitchell "be immediately relieved from duty as chief of Air Service, 1st Corps, and that he be sent to the United States."

Four days later I received a letter from Pershing's Chief of Staff, Gen. J. W. McAndrews. In it McAndrews said that Pershing had called Mitchell in, "talked quite plainly" to him, and "gave him to understand that he in no way approved or condoned his apparent insubordination, but that he would give him this opportunity to set himself right."

McAndrews continued: "The general wishes me to say to you that he wants you to meet Colonel Mitchell even more than halfway in this matter and to accept from Colonel Mitchell the assurances that he knows the latter is ready to give you. The fact that you have been his junior and are now his senior makes it possible for you to afford to do this."

Since Pershing had decided to keep Mitchell and let him stay under my command, I had to make the best of it. I called Mitchell in and told him and he generously agreed to let bygones be bygones. I pointed out that if his honest desire was to command a tactical air unit, I was going to give him every chance to prove what he could do.

Banking on Mitchell's General Staff experience, rather than on his air experience, I immediately authorized plans for the creation of a separate tactical unit which I designated the 1st Air Brigade. I assigned all units then under my command to this unit, and assigned Mitchell to duty as brigade commander.

A few days afterward the movement of American troops from the relatively quiet Toul sector to the Chateau-Thierry area began. I told Mitchell to proceed to Chateau-Thierry, take station with his units, and supervise their service with the commander of American troops in the area. It wasn't long before I started getting reports back that Mitchell had clashed with the other Air Service officers assigned to the 1st Corps headquarters because he was trying to take charge of all flying units in the vicinity, including those assigned to other officers.

I immediately investigated and composed a memo to Mitchell outlining the exact extent of his authority which General Craig signed as Chief of Staff. Characteristically, Mitchell replied with his theories on how aviation should be employed when working with ground forces, and there was a great deal of logic in his argument. On July 9 he sent me a memo explaining what he had done to get things under way and I was encouraged. If he had done and was doing what he noted in his memo, I was satisfied. There was no bitterness in his report and I thought maybe he would be a top-flight combat leader after all. If so, I was certainly going to back him to the hilt.

Although I still got reports of friction between Mitchell and the other Air Service commanders, they seemed to boil down to one fault. Mitchell was giving orders directly to his subordinate units without following routine channels. I investigated the matter thoroughly and concluded that in view of the military emergency at that time I would probably have taken the same action. It was then that I did some serious wrestling with my conscience. Maybe Mitchell is really a military genius, I told myself. Maybe he is one of those rare men in military history who, when given free rein and the weapons to fight with, can win battles. Perhaps my conflict with him was personal and that my attitude toward the man was really a reflection of my personal desire to lead America's combat air arm in battle. I found it hard to reconcile his insubordinate conduct but there was too much at stake. We needed men who could win battles. I decided Mitchell should have his chance to go down in history.

On July 25 I visited General Pershing and recommended that Mitchell have his old job back as chief of Air Service, 1st Army. I followed up the visit with a formal letter and cited Mitchell's most efficient service during the previous month, noting that his supervision over the air units in the Chateau-Thierry battle had "resulted in a minimum loss of life, a maximum effective use of material available, and a high fighting spirit of morale which will be most beneficial in establishing the standard of efficiency for all new Air Service units now organizing and to be organized in the future."

At our meeting, Pershing asked me what I thought I should do if Mitchell took over my job. I told him that I thought I could best serve as General Patrick's chief of Air Service Operations, where I could exercise supervision over all technical and tactical training of American air units in France. Pershing was obviously fed up with a situation where one officer could cause so much friction, but I told him I had three reasons for my recommendation: First, in spite of our personal differences, I recognized Mitchell's good qualities when properly applied; second, I wanted to prove that, for my Commander in Chief, the service, and the country, I could subordinate my personal desires to lead combat air units; and third, I saw that the construction of airdromes and depots, the training of new pilots, and the supply system were all bogging down. I still felt directly responsible since I had gotten all these programs started and General Patrick was stuck with making them work without really understanding the technical factors involved.

Pershing approved my recommendations, orders were promptly issued, and I crossed my fingers. Mitchell would either end up the war a hero if I were right, or in a strait jacket if I were not. Historians concede him the status of the former but they also had to record later that his disloyalty to his superiors, his insubordinate manner, and his vitriolic criticisms of people in high places would ultimately lead to a court-martial conviction. To this day I have never been sure that I did the right thing. I was gratified by his later accomplishments but I have never forgiven him for his denouncement of my "carpetbaggers" and for spreading the story that Pershing had fired me because of incompetence.

On April 14, 1918, a year and a week after war had been declared, two American Air Service pilots—Lieuts. Alan F. Winslow and Douglas Campbell—scored the first victories over an enemy in actual air combat. The next day Maj. Ralph Royce made the first American reconnaissance flights over the enemy lines. On June 12 the first air raid by an American bomb squadron was made on Dommary-Baroncourt, near Metz. From that time on our air activity picked up as planes, pilots, and units were organized. By the middle of July there was no longer any doubt that American air forces were fighting the Germans at last and leaving their mark.

During the month of August, 1918, Air Service activities in the Toul sector expanded rapidly in anticipation of the coming Saint-Mihiel offensive. The result was an inevitable loss of coordination between the air units at the front and the supply and personnel replacement units in Paris and Tours. The only way I could see to control the situation was to establish a General Headquarters Reserve under my direct command to which would be assigned all tactical units in the Zone of Advance not yet committed to combat. General Patrick issued the order creating this force and also provided that all requests for pilots, observers and supplies had to go through me. This stopped Billy Mitchell and every other commander in the forward echelon from harassing rear-support troops and making "midnight requisitions."

The plan for the Saint-Mihiel offensive called for a simultaneous attack on the two sides of the triangle-shaped salient established by the Germans. One American corps was to attack the western force, delivering its main blow to the base of the triangle, while two corps attacked the southern force, with their main blow falling halfway between the base and the apex. Both forces were to meet near the center, pinching off all enemy forces caught in the tip. A French corps was to make a secondary attack against the tip and exploit whatever success the two main attacks achieved.

Supported by 1,500 planes (mostly British and French), the American attack opened on September 12. By the end of the first day the Americans reached all their objectives. The advance might have gone farther but Pershing called a halt because of the coining Meuse-Argonne offensive planned for later that month. The Saint-Mihiel action was the first major

World War I victory by the American Army under its own command.

The accomplishments of Mitchell and his staff during this operation were exemplary. The unit under his command was unique in air warfare because it was the first time that air units of all the Allies, except the English (who were cooperating), had banded together in a common effort. It also marked the largest concentration of aviation placed under the command of one man (Pershing) up to this time. Of the 49 aero squadrons which took part in this offensive, 29 were American units.

Mitchell's staff devised an excellent plan divided into three phases. The first phase was the preparation for the attack; the second was to cover the ground attack to the first objectives; third was the exploitation phase during which the enemy was to be pressed vigorously on the ground and in the air.

The first phase was successful in that a total of 1,476 Allied planes were made available to Mitchell as opposed to 295 enemy planes backing up the enemy. Mitchell moved his headquarters from La Ferte-sous-Jourre to Ligny-en-Barrois. French airdromes were assigned to him and squadrons were moved into the Saint-Mihiel sector. Oblique photographs were made of the area immediately in front of our lines and American observers were made familiar with the front by making flights over it.

At 1 A.M. on the morning of September 12 an intensive artillery barrage began. The Air Service units were ready to fly at dawn but a downpour of rain, low clouds, and fog kept them on the ground. A few planes got off the ground later in the day and brought back the news that the Germans were withdrawing in great haste. The weather was just as bad on the thirteenth and I could well imagine Mitchell pacing back and forth swearing at the elements for being so unkind to him.

The fourteenth brought ideal weather conditions and missions of every type were dispatched during all the hours of daylight—observation infantry; contact patrols, photographic, pursuit, and bombardment. The Germans fought defensively in the air. They had to. But bad weather again hampered operations on the fifteenth.

The American air hero that emerged from the Saint-Mihiel operation has become a legend. His name was Lieut. Frank Luke of the 27th Aero Squadron. On the evening of the sixteenth he called the attention of the members of his squadron to two German balloons flying just behind the enemy front lines. Matter-of-factly, he told his buddies that if they would watch those balloons they would see the first one set on fire at exactly 7:03 and the second at 7:07. Like Babe Ruth pointing out where he would hit a home run, Luke made good his boast. Two days later he shot down two more balloons and three enemy planes in less than twenty minutes without getting a scratch.

The air operation at Saint-Mihiel was encouraging but has been exaggerated over the years, mostly by Mitchell himself and writers who have taken his "history" as gospel. He claimed in his memoirs that before the battle he "contemplated not only facilitating the advance of the ground troops but also spreading fear and consternation into the lines of communications, the enemy's supply system, and the cities and towns behind it which supplied our foe with the weapons of war." These were grand-sounding words but actually the missions his units flew were in strict subordination to Pershing's needs on the ground. Mitchell never allowed his planes to operate on more than a 35-mile radius of action. His real achievement was in gaining air superiority over the weaker German air force covering their front lines. The secret lay in the massing of the 1,476 planes before the battle without the Germans finding out about it.

The American victory delighted Pershing and he wanted to reward everyone in sight. General Patrick asked me if I thought he should recommend Mitchell for promotion to temporary brigadier general and I concurred. In all fairness, I thought Mitchell had certainly carried out a tremendous operation in spite of his inability to get along with people. On October 1 Mitchell pinned on the star he had obviously wanted for so long.

The Meuse-Argonne offensive began on the morning of September 26. Again Mitchell's staff had planned well. The tactical situation varied from the Saint-Mihiel offensive in that it was a frontal attack against organized and strongly defended trench works. Our forces enjoyed the advantage of superiority in manpower, tanks, artillery, and aviation. For this attack we had 821 planes available, 584 of which were manned by Americans.

The first phase of the operation required the ground forces to penetrate two of three enemy defense lines. This was done by October 1. Between October 4 and the end of the month our forces slowly drove their way into the third German line. Gains were small but significant.

What was most significant from the airman's point of view was the price we had to pay for lack of good war training and war experience. Our bombardment and pursuit units, full of eager but green pilots, had high losses owing to enemy air action, defective equipment, and poor defensive tactics. Even so, the total effect was much more successful than at Saint-Mihiel.

In spite of what Mitchell and air-power publicists have claimed, I do not consider the World War I record of our Air Service impressive. About 650 airmen claimed destruction of 491 enemy planes confirmed and 354 unconfirmed and 82 balloons shot down, of which 57 were confirmed. One hundred and fifty bombing attacks were carried out against targets as far as 160 miles behind the German lines. But the price of unpreparedness was high. On Armistice Day American units had 740 planes assigned but only 196 of them had been made in America. Five hundred and eighty-six American air officers and men had been killed, wounded, or missing, many of whom might have been saved if they had been required to wear parachutes as our balloonists did. Two hundred and eighty-nine planes and 48 balloons were lost, including 57 planes flown by American pilots assigned to British, French, and Italian air forces.

Looking back on the raw beginnings of the Air Service in France, I would never have bet that we would have amassed the air strength we had on Armistice Day. After the last gun was fired in anger, statisticians recorded that 1,402 pilots, 769 planes, and 252 balloon observers had been sent to the Zone of Advance before the end of hostilities. American aviators had flown 35,849 hours over the enemy lines. Our pursuit fliers had engaged in 2,100 dogfights and made 12,830 pursuit missions; 6,672 observation missions were flown and 17,845 photographs were taken of enemy positions.

The year 1918 has been called "The Year of the Sky Kings" for it saw the glorification of the pursuit pilot who became an "ace" by shooting down five or more enemy planes. The names of Rickenbacker (who had actually started the war as Townsend Dodd's chauffeur), Luke, Lufbery, Bishop, and Richthofen became more familiar to Americans than most of the generals of the war. Altogether 63 Americans qualified for the coveted title of "ace" by shooting down 462 of the 491 enemy planes officially confirmed.

On April 1, 1918, almost a whole year after war had been declared by the United States, there was but one Air Service tactical squadron operating on the front—the same 1st Aero Squadron that I had taken to Mexico in 1916. On the day hostilities ceased, there were 45 squadrons on the front, with 6 more squadrons in training behind the front getting ready to assume their places in the line.

If the war had continued there is no doubt that our combined Allied air power would have beaten Germany to her knees. In May 1918, the aviation committee of the Supreme War Council, upon which I served as General Pershing's representative, began to make comprehensive plans for combined strategic offensive operations in the spring of 1919. Based on our production estimates of Britain, France, Italy, and the United States, we foresaw that 2,500 pursuit and bombardment planes would be available for strategic operations and still leave hundreds of planes available for tactical operations with the ground forces.

We contemplated that each Allied air force would be

commanded by a chief of its own nationality under the direction of the general commanding the Army Group of the sector in which it was stationed but under the over-all command of the Commander in Chief of the Allied armies, General Foch. Although the plans for strategic air operations were fully developed by June, 1918, their actual execution was only partially started prior to the cessation of hostilities. Bombing operations conducted by the British Independent Air Force in the eastern part of France from July to November 1918, may be classified as a part of the strategic air operations contemplated for the fall and winter of 1918 and the spring of 1919.

Although Billy Mitchell has claimed personal credit or alluded that he deserves a place in history for the general concept of strategic operations, the record must be set straight. The concept must be credited to a number of men, such as Douhet, Caproni, Smuts, Trenchard, as well as my own dedicated staff who served with me on the Supreme War Council. Mitchell had no part in the development of our master strategic bombardment plans although he certainly became a renowned proponent of them after the war.

It is interesting to note the following preface to Mitchell's memoirs which shows how his writing and thinking have distorted history:

> It was my privilege to lay the foundations for the participation of our country in the air operations of the last war. This foundation, well laid, was torn to pieces by scheming politicians, both in the army and out of it, which delayed the creation of our air force by many months, causing the wasting of hundreds of millions of dollars and the lives of great numbers of our brave men. If it had not been for the arrangements we made with foreign powers to arm, equip, and instruct our men, we would have been perfectly helpless in the last contest. It was not Congress that was to blame for this, nor was it the people; it was the regular Army and Navy. The attempt by the personnel of these organizations to run them as closed corporations and to furnish misleading information to Congress and the people about the actual condition of our defense is what rendered us helpless in this war and in the wars that have preceded it....*

What a pitiful summing up of our air operations at home and overseas during World War I. What a pitiful effort to blame all of our failures on the regular Army and Navy, and what an astounding statement to say that "it was my privilege to lay the foundations for the participation of our country in the air operations of the last war."

It goes without saying that Mitchell was not one for giving credit where credit was due. His own lack of knowledge and inability at home and overseas to get results as an Air Staff officer did as much to delay our getting started overseas as were our difficulties with the French over engines, with the British and Italians over planes, and our own ground forces for construction and supplies. I honestly believe that my "carpetbaggers" could have accomplished much more during our first six months in France if we had not had so many problems owing directly to Mitchell's ego and personality.

When the hostilities were over, General Pershing asked me to make a report for historical purposes which detailed American air participation in the war. I prepared a 50,000-word document which pointed out our failures as well as our successes. I pulled no punches in regard to personalities, nationalities, or shortcomings. I believed then and I do now that our aviation progress depends on an intensive study of our failures—not our successes—and that these failures are of military value through the lessons learned. I think it is the duty of every officer in all of our services to profit by the bitter lessons of war and to let no obstacle stand in his way to correct what he knows to be wrong.

The report I rendered acknowledged that the Air Service, like every other branch of the Army, learned many valuable lessons in organization, supply, maintenance, and operations. Errors were committed both at home and in the field. Many of the major errors could be traced to conditions over which the Air Service had no control or were chargeable to superior authority. Other errors were

* Mitchell *op. cit.*, p. 9.

committed both at home and abroad over which the Air Service did have control. I wanted to point out all of these things for the future good of American military aviation and I did, but my frankness was not appreciated by my colleagues in Washington, nor by Billy Mitchell, who had emerged from the war as a national hero. The colorful, flamboyant, handsome son of a senator was photogenic and good newspaper copy because of his willingness to condemn people in high places.

The press and the public were led to believe he had fought the war in the air by himself. The irony of it all was that he was actually a dangerous pilot because of his lack of training, had never fired a shot at the enemy, and had done none of the master planning for the operations he supervised. Yet he received the Distinguished Service Medal, the French Legion of Honor, and the Croix de Guerre, and came home a general. I did not feel any bitterness, however, in spite of what the reader may think at this point. I had done the best I could to support my Commander in Chief, General Pershing. In his own memoirs of the war, he noted that "Foulois' desire to secure general cooperation made him a valuable assistant and but for his experience and his efforts we might not have avoided so many of the pitfalls that lay in our way."*

That was the only acclaim I wanted. I value those words more highly than medals, press notices, or ticker-tape parades. I had pleased my commander and that was all I was supposed to do.

* John J. Pershing, *My Experiences in the World War,* Frederick A. Stokes Co., New York, 1931, vol. II, p. 30.

Chapter Eleven

German Genius and Mitchell Goes Juramentado

There was a tremendous mental letdown for all of us after the Armistice was declared. We were glad, of course, that the fighting was over, but working and worrying seven days a week with the biggest problems of our lives to solve it wasn't easy to stop the pace so suddenly.

While the troops went home by the boatloads to their ticker-tape parades and the greatest military force our nation had ever seen deteriorated overnight, there was still a job to be done in Europe. The Armistice had provided that the Germans must surrender 5,000 cannons, 30,000 machine guns, 3,000 trench mortars, and 2,000 planes. A peace treaty had to be worked out by the Allied nations and submitted to their respective governments for approval.

To carry out the Armistice and provide the Armed Forces to enforce this peace, the United States agreed to join France and Great Britain in a military occupation of Germany. In December 1918, the United States 3rd Army crossed into Germany to occupy a small segment of territory along the Rhine River centering on Coblenz.

Since I had served on the Supreme War Council for General Pershing, I was asked to remain in France to work on the air terms of the Treaty of Versailles and see that American military and commercial air interests were protected. I worked with General Patrick, and we met daily with our Allied counterparts hammering out the words that might be acceptable to all the governments concerned in all their languages.

General Patrick and I, with a small staff, worked all winter with the air committee while other negotiations were going on at higher levels. The basics were negotiated between President Woodrow Wilson, Lloyd George, Orlando, and Georges Clemenceau, and then sifted downward to many working committees. President Wilson succeeded in having the Covenant of the League of Nations incorporated in the treaty which touched off one of the bitterest controversies in American political history. In spite of this, the treaty was handed to Count von Brockdorff-Rantzau, chief of the German plenipotentiaries, on May 7, 1919, the fourth anniversary of the sinking of the *Lusitania.*

Although President Wilson signed the treaty in the Versailles Palace Hall of Mirrors for the United States, it could not be binding until it was ratified by the Senate. He presented it to the Senate in July 1919, only to have it soundly rejected in November. It was introduced the following February but again rejected with the greatest opposition caused by the League of Nations Covenant. An overwhelming victory for the Republicans that fall proved that the American people were unwilling to enter into an "entangling alliance" with foreign powers, especially in Europe. Believed safe behind the oceans, the American people adopted the ancient policy of "isolationism."

We felt reasonably sure in Paris that the treaty would not be ratified from the reactions in the American Press. If it had been, I was slated to proceed to Germany and serve as the American air officer on the Inter-Allied Commission of Control. Instead, I received orders to return to the States in June 1919. Upon arrival, I was stripped of my star as a brigadier general and reverted to my permanent rank of captain in the Infantry and my temporary rank of major in the Air Service.

Just as my meteoric rise in rank had surprised me, so did my sudden fall. The demotion had come about through an order from Gen. Peyton C. March, Army Chief of Staff, who arbitrarily reduced hundreds of officers to their old prewar ranks. His main purpose, so the rumor went, was to cut Pershing down to his old peacetime rank where March thought he belonged. The only way he could do it was to order the dissolution of temporary ranks on a selective basis so it wouldn't seem too obvious.

The reduction would not have been so hard to take if it had been universal, but it wasn't. Those high-ranking officers that March liked kept their ranks, and they, in turn, allowed their friends and favored subordinates to keep theirs. One of March's favorites was none other than Billy Mitchell, who kept his star as a brigadier general and was made assistant director of the Air Service under Maj. Gen. Charles T. Menoher, an artillery officer. Mitchell thus became my superior and our rank positions were again reversed. Although he had the office next to mine and I saw him often, he never spoke to me, and I found no reason to report to him in regard to my duties.

I was assigned to duty in the office of the director of the Air Service as assistant and later chief of the Liquidation Division. It was the function of our staff to settle all government contracts for air matériel that had been in effect at the time of the Armistice. All contractors had been notified to cease production as of November, 1918, and submit their claims. By the time I arrived, paper was overflowing the small office we had been assigned and I wondered how long it was going to take to complete the job.

While assigned to the Liquidation Division I received notice that I was going to be asked to appear before the Senate Military Affairs Subcommittee concerning a bill "creating a Department of Aeronautics… and Administration of a United States Air Force, creating the United States Air Reserve Force, and providing for the development of civil and commercial aviation."

On October 16, 1919, I drove to Capitol Hill and faced five senators for two hours. I answered their probing questions and then left a 30,000-word statement with them for the record. I covered past and present legislation and let go with both barrels on a number of subjects ranging from the Army's weak plans to build up the Air Service to the Navy's plans to tear it down. I recounted my difficulties in trying to convince the Army's diehards of the importance of the airplane to the military, the potential it had to American commerce, and the penalties the Air Service paid for not being able to arouse the Congress and the people in time to join the war in Europe with an air force worthy of the name.

I took particular issue with the almost total disintegration after the war of the air arm that had been built up at such great cost, although too late to be a decisive force in the war. I accused the Army General Staff of being "reactionary" and noted that there was not "one single case on record where a General Staff officer of the United States Army has contributed to the technical or mechanical development of aviation." I added that the General Staff "should have appreciated ten years ago the fact that the aeroplane was a thousand years behind the development of all other military weapons and that in order to bring it up to the same state of development with existing military weapons, the Air Service would have to be given every encouragement and every possible precedence over other branches of the service in order that it might take its proper place in line with the other combat services."

Once I had expressed myself, I criticized the procurement organization that the General Staff had proposed, the promotion system, the lack of interest in research and development, and then took on the United States Navy by citing the problems I had had in France when the Navy took priorities away from the Army for engine and plane procurement which upset our whole logistics plan.

This latter criticism raised some eyebrows and sent the admirals scurrying for cover because I had opened up a tender spot which was to fester for many years to come.

I revealed for the first time publicly how the Navy had not only tried to fight their own strategic war in Europe but had made independent arrangements with the Italians to purchase a considerable number of Caproni bombers to the disadvantage of the Army order that had been placed eight months previously. I also said that Maj. Fiorello D. La Guardia, our chief air officer in Italy, was able to document my claims because he had kept me constantly advised of the Navy's plans.

I next addressed myself to the question of what the United States was doing to compete with European countries for the commercial supremacy of the air:

> The United States at present is doing nothing from a material standpoint, but it has done a lot of serious thinking about aviation in the past, and is still thinking. It is thinking right at this minute that after appropriating approximately $1 billion for the development of aviation and an aircraft industry for the prime purpose of helping to lick the Hun that there should be something left over after the war to devote to the peacetime use of aircraft. I have considerable faith in the thinking capacity of the American public, and am of the opinion that the public does not place the blame for the present deplorable state of aviation and the aircraft industry in general upon the aviators. I am further of the opinion that they will demand a new deal in our aviation game….

I reserved the last part of my testimony to attack the Assistant Secretary of the Navy, who had appeared on the Hill, regarding the Navy's views about a separate Department of Aeronautics. This man was rich, influential, and pro-Navy all the way. He had said that "not only the Navy Department officially, but the entire naval service, is absolutely opposed to the creation of another and separate branch of national defense." Being opposed to his views had a special significance later. His name was Franklin D. Roosevelt.

I attacked Roosevelt's arguments from his opening statement to the questions he answered following his prepared testimony. He was so wrong in his facts and so biased in his opinions that it was easy to shoot holes in his reasoning. I criticized him indirectly by saying that "failure on the part of our naval authorities at home to present to the War Department a complete analysis of the relative value, use, and needs of naval aviation as compared to the more important value, use, and needs of aviation with the Allied armies in the field was, in my opinion, a grave strategical mistake, a mistake which not only came very close to bringing disastrous results upon the British air program but directly interfered with the development of the Air Service, AEF."

I closed my testimony, urging that all air activities of the United States should be efficiently combined and economically controlled and administered. In my final statement, I said that "the aviation issue is squarely up to the Congress, and the responsibility of its future efficient development rests solely in your hands."

The air around the War Department was very chilly when my testimony was released but, since I had let General Menoher, the director of the Air Service, read it before I had gone to the Hill, I couldn't be accused of going behind anyone's back. Having alienated the affections of the entire General Staff, Secretary Roosevelt, and the Navy, and top industrialists whose claims I had whittled down in the Liquidation Division, I had the feeling that there were many people who would be happier if I left Washington.

In the spring of 1920 I heard that a military attaché was needed in Germany who had practical knowledge and experience in both lighter- and heavier-than-air machines. There was not an American officer in Europe who could qualify, so I volunteered for the job. Since I had made so many officers of the General Staff unhappy, I figured I wouldn't have any trouble getting the assignment. However, I was wrong. General March, Chief of Staff, personally turned my nomination down, stating that "the work to which Major Foulois' detail was recommended could be performed by officers now on duty in Europe." When I heard this, I appealed to General Menoher to make a reclama pointing out that the Allies were putting the air terms of the treaty of peace into effect through the Aeronautical Inter-Allied Commission of Control. Since the treaty had not been ratified by the Senate, the United States had no representation on the commission. The net result was that the United States was getting nothing in the way of military, technical, and industrial data from the Germans. They had been ahead of us technologically throughout the war and we had never caught up. If we didn't send a man to Germany who knew valuable data when he saw it, we would be missing a golden opportunity which would hamper American aviation progress for decades.

My appeal was successful. I received orders to proceed to Berlin and arrived there in May 1920. Technically, since no peace treaty had been ratified, the United States was still at war with Germany. All official German sources of air information were available only to the representatives of the countries which had fought Germany during the war. However, since the United States had not signed the Treaty of Versailles, the English, French, and Italians would not give me access to any air information they had acquired. I even had difficulty getting to speak to the chief of the commission, General Masterman of the RAF. When I did, I was politely but firmly told that my government would get nothing through his office unless and until the United States Senate ratified the Treaty of Versailles.

This rebuff concerned me, but I had a job to do for my government and I felt there were other ways to get information. I found out that the place in Berlin where many of the German Luftwaffe pilots congregated was the bar of the Adlon Hotel. For the good of my country, I made that bar my headquarters for almost a year and found there the best sources of aviation information in all Germany.

Slowly but surely my daily appearances at the bar paid off. The former Luftwaffe pilots were hungry. I bought them meals. They were thirsty, and the German beer was bad. I introduced them to good Allied whisky. They were broke and I tided them over "until better times." Their tongues gradually loosened to repay me for my generosity.

The names of two of these pilots are familiar today: Ernst Udet and Hermann Goering. Both of them were typical of the others—strong, virile, jovial, and vain. Under the influence of good Scotch whisky they became talkative, and a philosophical pattern slowly emerged from our chats. In their minds the Luftwaffe had not been defeated. They told me that the only reason the war ended as it did was because the heads of the German government had been weaklings and the people were tired of war. If the war had continued, their great aerial fleets would have smothered Europe with a rain of bombs and brought the Allied ground armies to their knees. After this was done, they then had even greater airpower plans. In the Aero Club of Germany they had drawings of long-range bombers that would cruise to North America and bomb the principal American and Canadian cities into ashes. No, they said, the Luftwaffe had not been defeated and someday the world would realize this.

At first I laughed at their boasts, but they became too universal, too standardized. It was almost as if they were quoting a party line. When they got deep in their cups, I challenged them to give me some proof of their boasts if they didn't want me to laugh. They told me to be patient.

As the weeks went by, I found that I actually enjoyed a special status in Germany. Excluded from contacts with my Allied counterparts and being the only American airman in Berlin had their compensations. The State Department still forbade fraternization of their personnel with the Germans, but their policies didn't apply to me. I wore civilian clothes and could travel wherever I wanted to go. Therefore, I proceeded unhampered by any restrictions on my time or my movements.

Apparently, my Adlon bar buddies checked up on me because every once in a while they would drop some tidbit about my past experiences into our conversations which I had not given them. They asked about the Wright brothers often, and laughed at how stupid the American people were not to recognize their achievement and learn how really to use the flying machine. "We will teach them, eh, Udet?" Goering chuckled as he jabbed his elbow into Udet's ribs. Udet, unsmiling and with a faraway look in his eyes, would always nod affirmatively.

It took about six months for me to gain the complete

confidence of Goering and his mates. And when I did, I didn't ask directly for information. Instead, I asked for industrial aeronautical contacts to whom I could talk. My timing was right because the activities of the various Allied Control Commissions were extending into every corner of German industry. Every German factory or business was being searched for arms, munitions, planes, etc. The members of these commissions, I discovered, were mostly businessmen in uniform who were really looking for German inventions and exploitable ideas. What they thought they could use, they confiscated. What looked to be of no commercial value was destroyed. As a result, every foreigner was suspect. However, I was not included in the wholesale condemnation of non-Germans. My grateful drinking pals nominated me for membership in the two top aviation organizations in Germany: the German Aeronautical Scientific Society and the Aero Club of Germany.*

I attended every lecture, exhibit, and function of these two organizations and soon found that doors to industrial aircraft interests all over Germany were slowly opened to me. I was allowed to tour their factories, fly their planes, talk with their engineers, and see their models and blueprints. They were proud of their aeronautical achievements and they had a right to be. The war had not dampened their imagination or their enthusiasm. They showed me many things hidden below ground in tunnels and cellars which no other foreigner had seen. I was frankly amazed at their ingenuity and could not hide my surprise. They were still so far ahead of the rest of the world that I was genuinely shocked.

If the United States was going to reap any benefits from German research and advancements of the flying art, I felt that the only thing to do was to approach them openly and offer to pay for their inventions. I soon found that this is what they wanted me to do, but they would sell on only one condition—I was not to give any information or items to the other Allies.

From the moment I agreed to this condition the United States

* The Aeronautical Scientific Society was the equivalent to our own National Advisory Committee for Aeronautics (NACA) which is now the National Aeronautics and Space Administration (NASA).

leaped ahead two decades in aeronautical progress. Using military intelligence funds, I started shipping German studies, plans, blueprints, and reports home by the ream. I was taken into the innermost recesses of aircraft, armament, and instrument factories and shown exactly how far ahead the Germans were in aeronautics. They had been experimenting with new aircraft and weapon designs that were so far ahead of the times that it wasn't until more than forty years later that I fully realized how advanced they were. They had new bombs, torpedoes, bomb sights, machine guns, photographic equipment, engines, planes, propellers, radios, gasoline, lubricants, and instruments that no other non-German had seen or knew existed. They had experimented with helicopters, rockets, gliders, parachutes, fuels, and high-altitude flight. They were thinking of huge long-range planes, in-flight refueling, submersible airplanes, and intercontinental rocket flight. In fact, I saw nothing during the period 1920-1924 that did not eventually come into being at a later date.

The more I looked, the more I found available for purchase. But I also found that I had to maintain the confidence of the contacts I made or my access to their secrets would be cut off. Goering told me at one point that my status was endangered because I had been seen playing tennis with a British member of the Control Commission. From that little bit of information, I knew how fragile were my relations with the Germans and how scrupulous I had to be to prove that I would keep my word.

As the weeks passed, I started shipping all sorts of aeronautical material out of Germany to the States right under the noses of the Allies. At one point I had accumulated enough to fill a boxcar, but the problem was how to get it past the French inspectors at Coblenz. I asked that a young lieutenant, Ernest E. Harmon, be sent to me from England to act as a courier. I arranged through American Express to have the boxes loaded at night in the freight car on a siding and then attached as the last car of a long passenger train. Harmon rode "shotgun" inside the car with instructions to do everything in his power to keep the French from finding out what was inside those boxes.

Harmon told me later that the scheme worked nicely. The train was so long that when it got to Coblenz, the freight car was outside the station. When the French gendarmes passed through the last passenger car checking passports, they thought they were through. As soon as they got off, the train started up rapidly. By the time the chief inspector realized there was a baggage car on the end of the train, his shouts were too late. His men ran after the train hollering and waving their rifles in vain. Harmon leaned out a window and waved back. Weeks later the boxes arrived safely in the States and my reputation with the Germans for fair dealing remained intact.

My sources of technical information were not limited to my personal contacts. I studied the German language five nights a week and started a campaign of visiting bookstores and libraries. I found that many German scientists and military men had published their memoirs and reports. Their vanity was another gold mine. I hired translators and even put my wife to work culling out what was of possible value and sending it via diplomatic pouch to the States.

Another method I used to elicit information and inventions from the Germans was to provide United States material as barter. I wrote to fellow air officers in the States and got them to send me all kinds of technical bulletins, magazine articles, congressional handouts, and material from the Government Printing Office on air subjects. All of it was in the public domain in the States but the Germans had no organized way of getting it. Their thirst for knowing what was going on outside Germany was voracious and I did my best to quench it.

Still another technique I used was to arrange for the Germans to protect their inventions by letting them patent them in the United States. This appealed to them particularly because the confiscation policies of the Allies were robbing them blind.

Toward the end of my tour in Germany, my job was getting easy. I had made so many contacts that I felt that I knew more about what was going on than the German government and I may have been right. In looking back over my list of contacts which I provided to the Military Intelligence Division, it is interesting to see some of the names. Besides Goering and Udet, following is a list of other "names" that later became famous, one way or another:

> BARON VON CHEMNITZ, former German Minister to China, member of the Reichstag and close friend of the former German emperor.
> DR. HUGO JUNKERS, president of the Junkers Aircraft Co.
> DR. WALTER HUTH, president of the Albatross Airplane Co.
> DR. ADOLF K. ROHRBOCH, director of the Rohrboch Metal Airplane Co.
> HERR FERDINAND RASCH, former director of the Zeppelin Company and president of the German-Russian Air Traffic Co.
> GEHEIMRAT JOHANN SCHUTTE, president of the Schutte-Lanz Airship Co. and president of the leading technical aeronautical society of Germany.
> HERR G. KRUPP, then editor of the leading technical aviation magazine and a member of the famous Krupp family.
> DR. THEODOR VON KARMAN, professor of aeronautics and leading designer of helicopters.
> HERR G. DORNIER, engineer and designer of the Dornier metal planes.
> DR. HUGO ECKENER, director of the Zeppelin Airship Co.
> HERR EDMUND RUMPLER, president of the Rumpler Airplane works.
> HERR ERNST HEINKEL, director of the Heinkel Aircraft works.
> HERR A. W. FOKKER, head of the Fokker Aircraft Works in Holland.

In all, I listed 180 individuals as my principal sources of information. They represented the entire spectrum of the aeronautical sciences and were all in the top echelons of their professions.

When it came time to turn over my duties to my successor, Maj. Harold Geiger, in the spring of 1924, I made an extensive report to the chief of the Air Service, Gen. Mason M. Patrick, who had succeeded General Menoher. In the report I gave some advice to my successors. Perhaps the following might have some value to all United States military representatives today:

a. To deal with the [foreign] businessman exactly as one would deal with an American businessman; that is, openly and frankly.

b. To be absolutely pro-American at all times. In other words, to keep constantly in mind that a United States military attaché serves only his own country.

c. To refrain from taking sides in the international politics of the country and, whatever the personal sentiments or feelings for or against [that government], to refrain from expressing such sentiments publicly.

As I looked back over my four years as air attaché to Germany, I felt an honest sense of accomplishment. The Germans hated the British, French, and Italians for the plundering of their industries, but they had a more friendly attitude toward Americans. My policies of playing fair with our former enemies and paying for what we got netted millions of dollars' worth of advanced aeronautical data for the United States. I only hoped that it was being put to good use in America. To my eternal regret, it wasn't. The lack of an air intelligence collection system, inexperience on the part of the military intelligence officers in regard to aeronautics, and a lack of appreciation for the potential value of the fruits of German genius caused much of the material I sent to end up unopened in a warehouse and later sent to the trash heap.

When I enquired about this upon my return to the States, I received a classic reply from a high-ranking member of the General Staff. "Foulois," he said, "this is peacetime and the war has been over for more than five years. Most of that junk you sent was either so old or so farfetched that not even a museum would be interested in it. If we ever go to war again, none of *that* stuff will ever be of any value."

He was wrong. Dead wrong.

I had been promoted to temporary lieutenant colonel in February 1923. During my tour in Berlin I had received an interesting package via diplomatic pouch. It was the Distinguished Service Medal awarded to me for my service in France. It arrived with a curt request from Washington to sign and return the enclosed receipt.

I received orders to return home in April 1924, and report to the Command and General Staff School at Fort Leavenworth, which is what I had applied for. I was returning to a new Air Service and I wanted to be up to date on what all the services were doing before I joined it as an active participant.

While I had been away General Menoher had held the job as chief of the Air Service until 1921. Many fliers had had great doubts about him because he was not a pilot himself. These doubts became certainties in their minds when the Army Reorganization Act of 1920 cut the Air Service personnel to 1,514 officers and 16,000 enlisted men. Menoher had held a board, consisting of himself and four other artillery officers, and concluded that the Air Service was only an auxiliary to the ground forces—nothing more. His assistant, Billy Mitchell, disagreed and told the world about it. Menoher requested Mitchell be relieved as assistant chief. The request was denied so Menoher asked to be relieved as chief and was.

Knowing Billy Mitchell as I did, I'm sure he thought he was going to get Menoher's job and a second star. But his public pronouncements about the potential of air power, his insinuations that the War Department was being run by idiots, and his continual fights with the Navy did not bespeak the kind of man that should have the top job. If he talked about his superiors as the number-two man in the Air Service, what would he say and do if he were chief?

Instead of Mitchell, Maj. Gen. Mason M. Patrick was appointed chief in October 1921, for the second time. His job was to attempt to restore discipline to the Air Service. His biggest job, however, was to try to keep Mitchell from getting out of hand, yet give him just enough rein to prove what the airplane could do for the military. It was an almost impossible task; yet under Patrick's firm but fair leadership great strides were made.

The Air Service achievements during Patrick's regime included many "firsts" and records. Night cross-country flying was begun; the first spanning of the country in less than twenty-four hours was accomplished by Jimmy Doolittle flying a DH-4B; Lieuts. J. A.

Macready and O. G. Kelly set an endurance record in a Fokker T-2 of thirty-five hours, eighteen minutes; world speed marks were set and broken several times; aerial refueling was accomplished successfully; the nation was spanned nonstop by Macready and Kelly; and an altitude record of 31,540 feet was set by Macready and Lieut. A. W. Stevens on a flight during which an aerial photograph covering the greatest area of the earth's surface to date was obtained.

One of the most outstanding “firsts” accomplished took place in 1924. It was the first successful navigation of the globe by air, the first transpacific flight, and the first westbound air crossing of the North Atlantic, all achieved on one flight. A prodigious effort, American Air Service pilots succeeded where pilots from Great Britain, France, Italy, Portugal, and Argentina had failed.

I had a small part in the achievement while I was in Germany. I had been asked to provide maps, weather data, and information on airfields, fuel, and servicing available in Europe. I obtained this data from my Luftwaffe friends and my contacts in Denmark, Sweden, and Norway. Since Iceland and Greenland were Danish possessions, I also assisted in obtaining landing permission and arrangements for servicing from the Danes.

On April 6, 1924, four Douglas World Cruisers, led by Maj. Frederick L. Martin and each manned by a pilot and mechanic, took off from a lake near Seattle and headed north to Alaska. One hundred seventy-five days later two of the original planes returned to complete the first round-the-world flight in aviation history. The route, covering 26,350 miles, was completed in three hundred and sixty-three hours of flying time—a magnificent accomplishment in the face of great odds because of the weather and grave risks owing to the pitifully primitive ground facilities on the route.

A typical incident which shows the Patrick-Mitchell relationship during this period took place in October 1922, when Mitchell told a reporter that “I am arranging for a 'round-the-world' flight with a squadron of six planes.” When this quote hit the newspapers the deputy Chief of Staff of the Army called Patrick in and demanded to know what was going on since the flight had not yet been approved. Patrick's answer proved that he, not Mitchell, was in charge. "By *my* direction," he told the press, "one of the divisions in *my* office has been giving study to possible long flights."*

I have no quarrel about Mitchell's championing the need for air power before the American public. It was his methods and his lack of judgment about what he said that I deplored. His flair for the dramatic had reached an all-time high when Air Service planes sank the German dreadnought *Ostfriesland* in July, 1921. He had taken the Navy's water admirals on barehanded when he said that battleships were obsolete and that the outcome of any future war would be decided by superiority in air power. Secretary of the Navy Josephus Daniels challenged Mitchell's clairvoyance and made the most unfortunate statement ever made by a Navy official: "I would be glad to stand bareheaded on the deck or at the wheel of any battleship while Mitchell tried to take a crack at me from the air. If he ever tries to aim bombs on the decks of naval vessels, he will be blown to atoms long before he gets close enough to drop salt on the tail of the Navy."

By the time I came home, Mitchell was riding the wave of controversy to the limit. Refusing to wear the regulation Army uniform and sporting a gold-headed swagger stick, his arrogance was sickening to many of us who felt there were more intelligent ways to accomplish the Air Service improvement objectives. As I began my course at Fort Leavenworth, the Mitchell controversy was coming to a boil. There was nothing General Patrick or anyone else could do to change Mitchell's course. He had become a fanatic much in the way that the Moros were in the Philippines. He had become a juramentado and was ready to run amok!

* Alfred F. Hurley, *Crusader for Air Power,* Watts, New York, 1964, p. 79.

Chapter Twelve

The Time for Patience and Progress

My assignment to the Command and General Staff School at Fort Leavenworth in 1924-1925 came at an interesting time in the history of the Air Force. The successful completion of the round-the-world flight focused attention on military aviation as no other single air feat had done. It was a sign of things to come, and I hoped I would have a hand in building what we needed in the way of military air power in this country. I had reason to think I would.

En route to Leavenworth I had stopped off in Washington to pay my respects to General Patrick and he asked me if I would like to be considered for the job as assistant chief of the Air Service when I completed school. "To be perfectly fair with you, I'm also considering Jim Fechet," he told me.

"Frankly, General, I don't think I'm ready for the job," I replied. "I've been away from the States too long and haven't commanded a flying unit for a long time. Besides, Jim Fechet is older and a good man. I think he should get it and that I should take a tactical command job when I finish school."

General Patrick agreed with my reasoning, and Fechet was appointed to the job in the spring of 1925. Although I appreciated being asked, the significance of the conversation escaped me at the time. My basic branch was still infantry and I was being carried on the infantry promotion list; therefore, I needed the Leavenworth school behind me—just in case the General Staff should decide that I must finally have to join troops in the field because I had not served with my basic branch since 1908. What had been Patrick's main concern was to provide a continuity of experience—practical experience—if the Air Service was to survive. He did not believe it could survive under Mitchell, and Mitchell was now giving him king-sized headaches as assistant chief of the Air Service.

I proceeded to Leavenworth and dived into my schoolwork. Still I couldn't help watching aviation developments. In the spring of 1924 the House of Representatives appointed a special committee to study alleged monopolistic practices in the aviation industry. Led by Congressman Florian Lampert, it expanded into a drawn-out enquiry into both military and civilian aviation. It had been preceded by a War Department enquiry into Army aviation two years earlier headed by Maj. Gen. William Lassiter. The Lassiter Board's purpose had been "to consider in all details a plan of war organization for the Air Service and to make recommendations to the War Department in the premises." Serving on the board were two old friends and fellow pilots, Frank Lahm and Herbert Dargue. This group, in spite of the foot dragging of Brig. Gen. Hugh Drum, came up with the first real plan for the future of Army aviation. They concluded that "our peacetime Air Service (must) have a strength and an organization permitting rapid expansion to meet the first requirements of a war and then be capable of steady expansion to meet the ultimate requirements of the war." Recognizing that duplications should be avoided, the board recommended a joint aircraft development effort with the Navy.

The Lassiter Board, called at the request of General Patrick, had pointed out that "unless steps are taken to improve conditions in the Air Service it will, in effect, be practically demobilized at an early date." The final report with its recommendations was forwarded to Secretary of War John W. Weeks in April 1923. It represented a ten-year program with a total cost that might have gone over $495 million. Unfortunately, the program never got airborne. President Harding's death brought Calvin Coolidge into the White House with a commitment to adhere to a rigid economy program which did not include aviation improvements. Coolidge's lack of interest was reflected in a story that circulated then to the effect that he asked Secretary Weeks, "What's all this talk about lots of airplanes? Why not buy one airplane and let the Army pilots take turns flying it?"

The Lampert Committee investigation was the aftermath of the steadily deteriorating military air situation. The Lassiter program was the closest thing to a sensible plan that the Air Service had known since World War I and the Lampert Committee wanted to know why it had not been adopted. Into the middle of the agitation stepped Billy Mitchell who had just returned from a honeymoon tour of the Far East with his new bride. Armed with theories and loaded with "facts" as a result of his trip, he began a speaking and writing campaign that was calculated to bring the attention of the people to the pitiful state of their Army aviation. It did.

In spite of a directive from the Secretary of War that Mitchell was to clear all speeches and articles with the War Department, he circumvented the directive and went his merry, fanatic way. He wrote a series of five articles for the *Saturday Evening Post* and harped on "his" theories of strategic bombardment which my staff and I had started to put into effect in France in 1918. Giving no credit to anyone else, the concept of the use of air power had, in the public's mind become all Mitchell's. Of course, all of us who flew agreed with his basic thesis that we needed a military air force capable of sustained operations against an enemy but we could not swallow his accusations that no one in the War or Navy Departments was doing anything about the sad situation. It was clear that Mitchell's sole objective was controversy and that he had gone far beyond the point of no return in the fall of 1924.

It was inevitable that he would be called to appear before the Lampert Committee and another set of hearings on a bill calling for a unified Air Service called by Congressman John F. Curry of California. In spite of the fact that the War Department now had a plan as a result of the Lassiter Board and the Navy had a viable aviation program under Adm. William A. Moffett, Mitchell charged both departments with gross neglect. When given the opportunity to back up his allegations with proof, he couldn't do it. He made statements before both committees that severely damaged his credibility, and Secretary Weeks told him that his term as assistant chief of Air Service would not be renewed and he would not become its next chief. Weeks wrote to President Coolidge saying that Mitchell's "whole course had been so lawless, so contrary to the building up of an efficient organization, so lacking in reasonable teamwork, so indicative of a personal desire for publicity at the expense of everyone with whom he associated, that his actions render him unfit for a high administrative post such as he now occupies."*

In March 1925, Mitchell was transferred to San Antonio and reverted to the grade of colonel. He could not be silenced and kept up his fanatical tirades in the press. In Washington, the real hero of air power, General Patrick, was calmly and efficiently winning the battle for the Air Service the only way a military man should—through proper channels and supported with overwhelming ammunition in the way of facts and clear logic. In December, Patrick asked that the War Department immediately implement the Lassiter program and create an Army Air Corps which would be similar to the semi-independent status of the Marine Corps within the Navy. It was a brilliant move and represented the next logical step toward a separate air force.

While at his "isolation" post in San Antonio, Mitchell published a very poor book entitled *Winged Defense.* Shortly afterward two tragic incidents occurred which again vaulted him into headlines. First was the disappearance of a Navy PN-9 plane on the first attempt to fly from the West coast to Hawaii. Then the Navy's dirigible *Shenandoah* crashed in Ohio. Mitchell called a press conference in San Antonio and handed them a nine-page mimeographed indictment of "the incompetency, criminal negligence, and almost treasonable administration of the Navy Defense by the Navy and War Departments." When I read this in my newspapers at Leavenworth, I was shocked. A civilian could say things like that but not an officer on active

* Letter, Weeks to Coolidge, March 4, 1925; Lampert Committee, 3036, 3039, 3040.

duty who had obligated himself by his commissioning oath to an unswerving course of loyalty to his civilian and military superiors.

The reaction of President Coolidge to this new insubordination was calculated to satisfy those who sided with Mitchell and yet preserve law and order within the military. He formed an investigating board headed by Dwight W. Morrow, a lifelong friend and partner in the firm of J. P. Morgan. A few days later he preferred court-martial charges against Mitchell after making sure that the customary investigation had been conducted. However, the Morrow Board investigation was to begin before the court-martial and was to make a study of the best means of developing and applying aircraft in national defense. The purpose apparently, was to prevent Mitchell from making the court-martial appear to be a lynching party taking an innocent man to his death.

The Morrow Board conducted bearings for about four weeks during September and October 1925. Approximately one hundred witnesses were heard, including myself, more than half of whom were fliers. The board found that there was a great conflict in the testimony, but in all the confusion of widely varying opinion concluded that in determining an aviation policy for the United States, military and civilian aviation should be distinctly separate. It also found that the United States was in no danger of air attack at that time from any potential enemy of menacing strength. The board did not recommend a Department of National Defense, either as comprising the Army and Navy, or three coordinate departments comprising the Army, Navy, and Air. Neither was the board in favor of a separate department for Air coordinating the Army and Navy Departments.

Mitchell had been asked to appear and spent his first four hours reading from his book *Winged Defense.* Morrow let Mitchell exhaust himself reading, and when Billy finally ran out of gas, adjourned without challenging him on any of his statements. The next day Mitchell did a very bad job answering the board's questions. The board ended its public hearings and began preparing its report. The court-martial began on October 28, 1925.

There were twelve generals assigned to act as judges, including Douglas MacArthur. Three of them were successfully challenged off the court and the remaining nine rendered the historic "guilty" verdict of conduct prejudicial to "good order and military discipline [and] conduct of a nature to bring discredit upon the military service."

There were very few regular Army officers who sympathized with Mitchell. We were glad that he had focused national attention on the tragic state of our air power but could not condone his methods. General Pershing referred to his conduct as the result of an insidious "Bolshevik bug" which had to be eradicated for the good of the Army.* Most of us agreed. The consensus of those who had served with Mitchell and had had to live with his slander and invective was that he had gotten exactly what he deserved. He was sentenced to five years' suspension from active duty without pay or allowances. President Coolidge changed the sentence to five years' suspension at half pay. However, Mitchell chose to resign in February 1926, and thus gave up all claims to any retirement benefits.

While all of these developments were going on at the national level, I studied them at leisure at Leavenworth. The Morrow Board and Lampert Committee recommendations and the Mitchell court-martial were milestones in the fight for air power. Numerous bills were introduced in the next session of the Congress. After extensive hearings, the bill which later became known as the Air Corps Act was finally passed and approved on July 2, 1926.

When I completed my course at Leavenworth, I got my wish to command a flying unit. Having been on the Detached Officers List since 1920, I had received no flight pay and the only flying I had done was with a few of my Luftwaffe friends in German planes, and a few flights at Leavenworth. I was as eager to get my hands on the controls of our new planes as a teenager approaching the driving age. My orders read that I was to assume command of

* Alfred F. Hurley, *Billy Mitchell,* Watts, New York, p. 104.

Mitchel Field, Long Island, New York, and also command the 9th Observation Group there."*

Mitchel Field had become widely known as the "showplace of Army aviation" by the time I arrived. Annual air races had been held on the field and every publicity stunt involving Army airplanes that could be imagined had been tried. I decided to change all that and put us back in the business of tactical aviation. It wasn't easy.

One of the first tasks handed to me was the planning for the 1925 air races. To accomplish my personal objective of building a tactical unit, I decided to make the 1925 races as military as possible. I drew up "Race Orders" and detailed responsibilities down to the last cook and carpenter just as if we were going on a military maneuver. The end result was a well-ordered and controlled International Air Meet for which we received many letters of appreciation. Better than that, I had formed the nucleus of a tactical organization that had produced something in the process and now had an *esprit de corps* that I could build on.

My next task was a more serious one. Mrs. Foulois, upon our arrival at Mitchel, had taken charge of the welfare work on the base. About a week after the races she told me she had just come from the hospital where she had talked to the wife of one of our noncommissioned officers who was confined to the base hospital for mental treatment. The distraught woman told Mrs. Foulois an incredible story that her husband was mixed up with a gang of rumrunners who were operating a fleet of motorboats along the Atlantic coast which were powered by Liberty engines supplied by her husband.

My wife repeated the story to me and I immediately had the base provost marshal locate the noncom and bring him to my office. At the same time I telephoned the Nassau County chief of police and district attorney and asked them to come to my office. My wife repeated the story in front of the suspect and the officials

* Contrary to popular misconception, the field was not named for Billy Mitchell. It had been named for John Purroy Mitchel, a former mayor of New York City who had been killed in Texas while in training for Army service.

and I reminded the astounded sergeant that the charges that could develop from such a story were serious and that anything he said might be used against him.

He denied the allegations, saying that his wife was unbalanced, and that no one should believe the ravings of a mental patient. However, the Nassau County officials took over, and their questioning was so blunt and so thorough that the sergeant finally confessed. He gave the names and addresses of ten civilians and another noncom with whom he had been dealing to get Liberty engines out of our warehouse and into the boats of the gangsters.

This enterprise had been a slick, profitable, and most successful operation for more than a year. Several hundred Liberty engines were stacked in the base warehouse and the loss of a dozen or so was never missed. Converted into a marine engine and put into streamlined boats, these engines saw service which their designers had never intended. No law-enforcing seacraft could catch them.

Both noncoms were court-martialed, but the ten gangsters were never caught as far as I know. There was some good developed as a result of the trial of the two noncoms, however. It marked the resumption of old-fashioned military discipline so essential to any military tactical unit.

In the months that followed I embarked on a campaign to remove Mitchel Field from the stunt and recreation business as painlessly as possible.

The requirement to assist in publicity stunts was a little difficult to shuck. Typical of these was the gimmick somebody dreamed up to have Babe Ruth try to catch a baseball dropped from an airplane. Duly covered by press, radio, and motion-picture reporters, the Babe got publicity all right. I had a pilot, Capt. Harold McClelland, go up with three baseballs and "bomb" the Babe with them from 250 feet. The first two balls knocked him flat, but he held onto the third one and gave it to me as a souvenir. The last I saw of the Babe that day he was slowly flexing his burning hand and trying to smile about it as he left in a big limousine.

In the period 1925-1927 I slowly tightened the reins and tried to whip the tactical group into shape. I made up training schedules for an entire calendar year, put the base on a rigid daily routine, required reserve officers to receive identical training as regulars, devised tactical problems for all the squadrons, and instituted a supply and maintenance system that would enable anyone to determine where we stood on shortages and overages and the flyability of our equipment.

In the spring of 1927 I got word that a reserve officer named Charles A. Lindbergh was going to attempt to fly the Atlantic from New York to Paris. He planned to do it in a single-engine plane—and alone. I knew he would need our help at Mitchel and he would get it. For a long time I had been trying to improve the weather information we had been receiving. I had arranged with the Navy, Coast Guard, Weather Bureau, and the Post Office Department Air-mail Service to exchange weather data via radio with my people. As a result, my pilots were better informed on what to expect along the Eastern seaboard than ever before. I volunteered my weather net to this young officer and he accepted. He made aviation history, and I think the data we provided helped a great deal. He later wrote me a letter of thanks saying "it was through you that I obtained the meteorological data… which enabled me to estimate the time of my departure."

After Lindbergh's historic flight, we provided marching troops for his triumphal ticker-tape parade in New York City. Three days later, as I sat in my office at Mitchel, a tall, thin figure dressed in a rumpled tuxedo and leather flying jacket appeared in the doorway and asked if he might speak to me. It was the same courteous, shy young man whose name "Lindy" had become the most famous nickname in the world.

Lindbergh confessed to me that he had just escaped from the clutches of his "managers" who had worn him out with press conferences, speeches, dinners, awards, and ceremonies. Would it be possible to get quarters on the base so he could get some rest?

I sympathized with this brave, unassuming lad and took him to my own quarters. I posted guards around the house and at his bedroom door with instructions to let no one disturb my guest. About two hours later two fast-talking young men arrived and announced they were trying to find Lindbergh. They were his managers, they said. He was due for an appearance and they knew he was on the base somewhere. They demanded that I lead them to him.

I listened to their demands while I stuffed my pipe. When I had taken a few infuriating minutes to light it and took a few puffs, I told them that they were talking as if they "owned" Mr. Lindbergh. "We do," they answered in unison.

"Nobody owns anybody in this country," I retorted. "If you think you do, I'll prove you don't." I pointed toward my quarters and said that Lindbergh was there as my guest. They started to bolt out the door when I called them back. "It is only fair to warn you that those men stationed around my quarters are armed," I said, "and have orders to prevent, by whatever means necessary, anyone seeking to enter that house. They are all sharpshooters."

Hesitating, the duo didn't know whether I was kidding or not. "Care to try?" I asked. Apparently, they didn't. The last I saw of them they were headed for the front gate wondering how they were going to solve the public-relations problem that I had foisted upon them by my unreasonable attitude. I have never discussed this incident with "Slim" Lindbergh but I'll bet he was grateful.

The Air Corps Act of 1926 gave us a tremendous shot in the arm but it did not last long. Although it changed the name of the Air Service to Air Corps, "thereby strengthening the conception of military aviation as an offensive, striking arm rather than an auxiliary service," nothing much changed in the general attitude of the War Department toward military aviation. Even the appointment of F. Trubee Davison, a pro-aviation spokesman, to the new position of Assistant Secretary of War for Air did not help very much.

Perhaps the most encouraging aspect of the act was the

authorization to carry out a five-year expansion program. We had suffered a chronic lack of funds in all the previous peacetime years and had never been able to build up to our authorized strength in men or planes. When the act was passed we had only about 60 pursuit and 169 observation planes but no bombardment or attack planes of any value. Only 115 training planes could be called modern; in fact, we were still using DH-4s made during World War I for training cadets.

Small and lasting gains were made during 1926-1927, however. The Air Corps Training Center was established at San Antonio under the command of Frank Lahm and included not only primary and advanced flying schools but also the School of Aviation Medicine. We also put our logistical organization on a businesslike basis by establishing the Matériel Division at McCook Field, Dayton, Ohio, in October, 1926. A year later it moved to Wright Field where the Air Force Logistics Command has its headquarters today.

More aviation records were set at the time the Air Corps was established which served to keep the Army air arm in the news. J. A. Macready, our foremost altitude pioneer, set a record of 38,704 feet in an XC05-A in January, 1926; in December Herb Dargue led a flight of five COA-1 amphibians on a tour of Central and South America. In February, 1927, as part of a continuing activity, the Air Corps made an aerial photographic survey of the east and west coasts of Florida for the United States Coast and Geodetic Survey. This was so successful that the program continued until 35,000 square miles in various parts of the States had been photographed for various government agencies.

Notable progress was also made in long-distance and endurance flying at this time. On June 28, 1927, Lieuts. Lester J. Maitland and Albert F. Hegenberger flew a Fokker trimotor, the *Bird of Paradise,* nonstop from Oakland, California, to Hawaii. In addition, civilian and Navy fliers set many records during this period and each one pushed the frontiers of flight back another notch. It was a necessary trial-and-error period which taught us that there was literally no limit to heavier-than-air flight except our own imagination.

Toward the end of 1927 General Patrick called me to Washington to tell me he was retiring and that Jim Fechet would be his successor. He asked me if I would consider taking the job of assistant chief of the Air Corps. I was delighted, but I hated to see General Patrick leave the service. He had held the job six years and had brought discipline and order out of the chaos that Mitchell had created by his constant insubordination.

Unfortunately, General Patrick is one of the least known of the chiefs of the Air Force yet did more for military flying during his tenure than people realize. His genius was organization. I had served with him in Cuba when he was the chief engineer with the Army of Cuban Pacification and in France. Although I did not always win my arguments with him, he always heard me out and then pointed out where my argument was weak. As a result, he was a great teacher and leader.

The reason why Patrick is not remembered is because he was overshadowed by "the martyr for air power," Billy Mitchell. But Patrick was the farsighted leader who appeared before committee after committee on Capitol Hill giving them calm appraisals of the future of aviation. While conducting his six-year fight for adequate legislation, he put his words into action within the limits that the law allowed. To stimulate aircraft production and airmindedness on the part of the public, he authorized scientifically valuable air exploits and records. He sought every opportunity to demonstrate the practicality of the airplane in peacetime such as forest fire patrols, rescue operations, crop dusting and aerial mapping. It was General Patrick who built up the Matériel Division in order to give the research, development, and logistics side of the Air Corps a chance to develop in an organized fashion. And it was General Patrick who saw that a foundation was laid to have training keep pace with technical advancement.*

General Patrick was the one who had asked for the Lassiter

* *The Air Forces Newsletter*, February, 1942.

Board to be appointed in 1923 which laid the proper groundwork for the Air Corps Act of 1926. Although this legislation was his swan song, it was a victory that should be remembered by military aviation historians. This man, who had learned to fly at age sixty because he wanted to know about the problems of fliers firsthand, opened the door to the future. It is unfortunate that Billy Mitchell has been allowed to claim so much for himself in the way of accomplishment during the half decade following World War I and that so many historians have believed him without checking their sources. Mitchell can certainly lay claim to the title of "spokesman for air power" but his personal accomplishments should be deeply questioned.

General Patrick retired in December, 1927, and I transferred to Washington. However, my transfer was easier to get than the job I had been sent to take. My previous tour in Washington as head of the Liquidation Division after World War I was remembered by the industrialists I had stopped from robbing the treasury. My appearance on the Hill in October, 1919, was vividly recalled by the Army General Staff and the few congressmen I had upset by my frank appraisal of military aviation and my attribution of its plight to their collective shortsightedness. The result was that confirmation of my appointment as brigadier general was withheld by the Senate Committee on Military Affairs. Senator Reed of Pennsylvania, the chairman, told the press he needed "more information concerning this officer's record." Later he told a reporter from the *New York World* that action on my appointment had been deferred "by request" but declined to say who made the request or the reason for it.

I began to feel like the illegitimate child at a family reunion. I took the oath of office as assistant chief as a lieutenant colonel on the same day that Jim Fechet moved up and General Patrick retired. One newspaper editorialized that "the story is that Colonel Foulois has enemies both in and outside the service, who, despite the fact that he has been an outstanding figure in aviation since the first Wright plane was purchased by the Army, might prevent his nomination, though they would not dare to come out in the open and attack him.... If the story is true, it bears out the theory held by many that this country, like others, has its own military clique, a closed corporation, which, because it has on hand a flag to wave when its actions are questioned, is immune to attack." I made no comments to the press and waited patiently. I had received wounds in other Potomac River battles and if the good representatives of the people in all their wisdom did not believe I was qualified to be a brigadier general again, there was nothing I could do about it except ask for a transfer out of Washington, so I wouldn't embarrass Jim Fechet or the man who would get the job in my place.

After about two weeks Jim Fechet called me into his office and he was all smiles. He came around from behind his desk and shook my hand warmly. "You're on the team, Benny," he said. "Now, let's get to work."

I never did find out who had held up my nomination or why. I didn't care particularly. I pinned on the star of a brigadier general for the second time and got busy.* There was much to do to get the five-year plan under way. One of General Patrick's last official acts had been to prepare estimates for the fiscal year 1928 as part of the five-year program. I could foresee another battle with the Army General Staff and I was right. I had studied previous budget estimates and saw many necessities that had not been included. Inexperienced budget makers who were not fliers had been too general in their fund requests. As a result, I suspect that higher-ups in the War Department chain of command and the Congress whacked off some of their gross estimates because they looked too large. I asked for more detailed figures from my own staff down to the last roll of toilet paper and airplane tire. There was a lot of balking at first, but the estimates got better. My reasoning was twofold: I wanted our staff to stop making wild guesses as to what was needed and I wanted those above the Air Corps with the power of cutting our requests to know what they were cutting.

As soon as the 1928 budget estimates were approved and forwarded, I sat down with Jim Fechet for a long talk. I had been

* I have never held the rank of colonel.

studying intelligence reports coming from Europe and I could see evidence that the Germans were building their air force in spite of that clause of the Versailles Treaty which stated: "The armed forces of Germany must not include any military or naval air force." I remembered the words of Goering and Udet in the early twenties when they pointed out that the Luftwaffe was not licked and that they would rebound with the greatest air power the world had ever seen. Although the regulation of civilian aircraft by the Inter-Allied Commission of Control had placed a restriction on commercial aircraft, they were developing larger planes with better, more powerful engines. "It would be no trick to design a transport for long-range flight," I told Fechet, "and make the design such that it could be converted into a bomber." I briefed him thoroughly on German basic air strategy that in 1920 was being projected to design and build medium-range aircraft for use against Europe and long-range aircraft for use overseas against Canada, the United States, and Latin America.

I ended my briefing stating that from now on it was my belief that we should concentrate on development of long-range aircraft, capable of taking the air war to Europe, instead of waiting for them to invade the Western Hemisphere.

Fechet, not having seen nor heard what I had in Germany, wasn't so alarmed by each new intelligence report that showed German advancements in "commercial" aviation as I was. The treaty had not stopped German imagination and genius from functioning. What worried me was what our intelligence did *not* know about.

"Jim, if you have no objections," I told Fechet, "I'm going to start planning beyond our five-year program and think in terms of long-range bombers and fighters and research into new weapons just as I'm sure the Germans are doing." Fechet had no objections, and I began a personal program of notemaking which I hoped someday would lead to a productive research and development effort. At every opportunity I inserted my ideas into war plans and procurement programs, and inch by inch I could see my ideas shaping up. Unfortunately, most of them had to be disguised by labeling them as "defensive" or "in support of Army and Navy forces" in order to get them by the General Staff. I particularly encouraged the full development of all-metal plane construction, pioneered by Hugo Junkers in Germany, and continued improvements in engines, propellers, retractable landing gear, bomb sights, gun turrets and instruments. All of these things had to improve simultaneously along with our training of personnel.

In the spring of 1929 I suggested to Jim Fechet that I needed more intimate knowledge of the organization and operations of the Matériel Division of the Air Corps, then located at Wright Field, Dayton, Ohio. This was the Air Corps unit then charged with the combined problems of research and development as well as all logistic problems incident to procurement, care, repair, and maintenance of aircraft matériel. At that time I was not satisfied with our progress in the development of long-range aircraft, engines, instruments, or fuels. In fact, I was not satisfied with our entire research and development program, which, though hampered by inadequate funds, I felt could be speeded up through closer cooperation between Air Corps Headquarters, the Matériel Division, and the unit commanders throughout the Air Corps. I suggested that Gen. William Gilmore, then commanding the Matériel Division, and myself exchange jobs for a year, in order to give me a better opportunity to study the detailed organization and daily operation of the Matériel Division.

This exchange of jobs was approved by Fechet, and I took over my new job as chief of the Matériel Division in September, 1929.

My next task at Wright Field in 1929 had chiefly to do with the Research and Development Section of the Matériel Division.

Throughout the fall, I devoted my time to the study of past experimental and service test contracts, as well as those then outstanding with aircraft matériel contractors and subcontractors. My studies disclosed two very important problems:

(1) Failure to use the practical experience of the users—the tactical and strategical units in the Air Corps—prior to awarding experimental contracts;

(2) The frequent use of "change orders" whenever some one of the scientists and/or technologists in the Matériel Division had some new ideas which they insisted should be incorporated in a live contract.

With reference to the first problem, in those days, as in the present day, research and development had as their primary objective the task of providing the users with the best aircraft matériel that human ingenuity could design. Throughout my then twenty-one years as an experimental test pilot it had always been my experience that many of the users, both enlisted men and commissioned officers, could always contribute valuable ideas in the field of research and development as based on their practical experience.

I issued a directive to the effect that hereafter no experimental or service test contract would be awarded until after interested squadron commanders and their engineering officers had been consulted and allowed to submit their own ideas based on their practical experience as users.

With reference to the second problem, the practice of issuing change orders had always resulted in not only delaying completion of a contract, but had always increased the original cost of the contract. Therefore, in order to prevent delay in the completion of such contracts and to prevent increasing the original cost, I issued another directive to the effect that hereafter no change order involving live contracts would be issued except in cases involving safety. And in such cases all change orders had to be submitted to the chief of the Air Corps for final action.

It was not enough merely to put in the paper work each year to attempt to justify what we needed in the way of equipment. We had to use what we had in simulated military exercises and then use the results as proof of our capabilities and evidence of our shortcomings. Accordingly, in the fall of 1930 after I returned to

Washington as Assistant Chief of the Air Corps, I had my staff devise an extensive plan for a realistic air exercise the following spring. My only guidance was to "think big" and "come up with something that will involve every plane in the inventory that will fly, including National Guard planes." Under the leadership of Lieut. Col. Frank M. Andrews, one of the ablest officers ever to wear wings, they went to work and came up with a plan that was the most extensive air operation ever made up to that time. They had taken the experience of the 1929 maneuvers in Ohio and the 1930 maneuvers in California, and came up with a plan to form an Air Division which in time of war could be moved as a unit from one place to another and consist of pursuit, bombardment, observation, attack, and utility planes. I approved this concept of mobility as the theme, and then gave them a list of questions which their plan had to answer if it were to do us any good in the future:

- Were our tables of organization and equipment and our supply and logistic schedules adequate?
- How many, and what spare parts had to be available at the various fields to ensure proper maintenance of the planes?
- Could an organization the size of our Air Division be handled with the staff assigned?
- Were present methods of communication adequate? What would be better?
- Could changes in operations or routes be made successfully after the units were in the air?
- How many planes could be successfully operated from one field?
- How long would it take for a large air unit to take off or land?
- How long does it take to service a large organization so they can resume operations?
- Could recent graduates of the advanced flying school or members of the graduating class take their places in a combat organization in time of war?

The staff went back to the drawing boards, and now plans took shape. I approved them and passed them to Jim Fechet. Orders for the assembly of planes, the placement of supplies and parts at selected fields began to flow out to the units. Press releases were prepared in which the rationale for the "Air Corps Coast Defense Exercises" was explained. At the beginning of May 1931, I held a press conference and announced that I was going to lead the maneuvers. I told them that "the Air Corps of *this* country must keep pace with that of the other leading world powers, both in the quality and quantity of their planes and pilots and in their staff as well. Without an Air Corps operation on a large scale, as we are planning this year, we will not be practiced in the handling of large numbers of planes and flying personnel. Without such practice we cannot say that we know how a wartime Air Force would be operated."

In spite of my best efforts to explain what we wanted to do, there were adverse reactions. People wrote to the newspapers and editors wrote editorials making somber predictions about the operation and lodged protests about the warlike character of the air armada. The overcautious complained that dozens of America's finest youth would surely die in accidents during the maneuvers. The pacifists protested that such an aerial show would only stir anxiety among the other nations of the world.

I ignored all of those against our plans and on May 10, our fields from New York to California launched their planes into the air, and headed them toward Dayton, Ohio. Six hundred and seventy-two planes concentrated at Wright Field on May 15, and made up the largest armada this nation had ever seen at one time in one piece of sky before or since. On the seventeenth, a Sunday, we had combat competition between two pursuit squadrons, and then the entire division took to the air to pass in review for the citizens of Dayton. The following Tuesday the division flew to Chicago and back in a 20-mile long cordon; on Thursday we flew to New York, where we made a simulated night bombing attack on New York City and gave demonstrations at Floyd Bennett Field. We flew to Boston on the twenty-fourth and from there staged simulated combat missions to Portland and Bangor, Maine. On the twenty-seventh we flew reviews over Trenton, Newark, and other New Jersey cities. On the last day, Memorial Day, we flew en masse to Washington for a display of aerial might that impressed even the most unsympathetic of our lawmakers.

The statistical results of the entire exercise were better than I had dared hope. There were no major accidents of any sort and the record of safety covering more than 500,000 plane miles was a tribute to my planning staff. Only those who fly can appreciate what I was going through mentally as I led those 672 planes in mass formation. Statisticians could, and did, calculate that the mere fact that there were so many planes in the air would lead to mid-air collisions and the number of landings and takeoffs at airports crowded with planes should increase the odds for accidents, but they just didn't happen. Our success caused Secretary Davison to say, "I doubt if any air maneuvers have ever been more productive than the present ones from the training and technical standpoint. They clearly place the United States in first place insofar as the operation of large units is concerned."

I don't know about the last sentence, but he was right about the first. We learned plenty from our failures, which is exactly what our purpose was. We didn't tell the press, but there were many mistakes made owing to inexperience and faulty equipment. In the weeks and months that followed I made sure that everyone knew what those mistakes were and that we were doing what we could to ensure they wouldn't happen again. At the War Department level, I used the lessons learned as a basis for bolder planning and another step to make the Air Corps a real fighting force.

The maneuvers received accolades in the press and from the President on down which were unprecedented. For my part in them, I was awarded the Mackay Trophy for "the most meritorious flight of the year." It was the only award I ever received for flying in my active military career.*

My feelings, however, were succinctly expressed in an editorial in the *Tucson Daily Citizen,* on June 3, 1931:

* The winner receives only the honor of having his name engraved on the trophy, now valued at $50,000 and kept under guard in the Pentagon. It has been awarded annually since 1911 and originated by Charles K. Mackay.

> .... It had been assumed that this greatest mobilization of the nation's air forces would be attended by loss of life. Actuaries on the basis of experience, had estimated the number who would be killed in accidents. That their estimate was wrong merely reflects the increased reliability of the airplane and the increased skill of the flier....

The fact that we had no one killed or injured throughout the entire period of the maneuver may be attributed to one major factor—fair weather. Had we encountered foul weather at any time with those big formations in the air, the score might have been different, because those were the days when training in the use of blind-flying instruments was in its infancy, and although some of our planes had been equipped with suitable blind-flying instruments since 1925-1926, many of our best pilots continued to depend on the old "iron compass"—the railroad—or were forced to land and wait for fair weather.

I had started blind-flying instrument training while on duty as chief of the Matériel Division in 1929, using the instruments then available. In the spring of 1930 the Matériel Division began to service test the artificial horizon and the gyro director developed by the Sperry Gyro Company. I had a set of these new gyro instruments installed in my airplane alongside of the old turn-and-bank and rate-of-climb instruments, and continued my blind-flying training, using both sets of instruments while at Wright Field, and later at Bolling up to May, 1931.

Later in the summer of 1931 some of my bomber boys wanted to carry out some bomb tests against a ship to see if 100- and 300-pound bombs could sink it. Having received the approval of the War Department General Staff, they arranged to have an old freighter, the SS *Mt. Shasta,* donated by the Shipping Board, towed out to sea off Norfolk. Herb Dargue was to lead a flight of nine bombers from Langley Field, locate the *Mt. Shasta,* and sink it.

After its sinking, a complete evaluation of the damage was to be made to determine the effectiveness of the bombs and exactly what caused it to sink. Full press coverage was to be given to the event.

I was against the idea because I saw it as nothing more than a re-opening of the 1923 controversy at a time when we needed more solid accomplishments to give us public support. However, as the project had been approved by the General Staff, Jim Fechet overruled me and took charge of the preparations. Press coverage was immediate, and ominous headlines began to appear. AIRPLANES VS BATTLESHIPS AGAIN; ARMY TO BOMB ANOTHER SHIP; OLD SHIP WILL BE TARGET FOR NINE ARMY PLANES. The slant these articles took was not exactly pro-Army Air Corps and I could see that if Dargue and his boys didn't sink that old tub, we were in for trouble. It seemed to me the tests could have been done without press coverage; but, once announced, the press leaped on the story and gave it undeserved attention.

On August 11 the Coast Guard cutter *Mascoutin* left the Old Point Comfort dock. Aboard were thirty newsreel cameramen and reporters. The *Mt. Shasta* was towed into position about 60 miles off the Virginia Capes. The planes took off from Langley in light rain and fog and started a search for the vessel. Four hours later Dargue had to give up the search. Low clouds, rain, high winds, and fog made the task impossible with no radio beacons to guide his pilots above the overcast. The only reason he had taken off in the first place was that he didn't want to disappoint the press. He later wished that he had.

This was all the press needed to give the Army a bad time. The headlines made fun of the "lost" pilots, and then, to make matters worse, the Assistant Secretary of the Navy for Aeronautics David S. Ingalls wrote the following letter to Secretary of War Patrick J. Hurley which he released simultaneously to the press:

> Dear Pat:
>
> I recall with interest an incident that occurred during the President's trip to the Virgin Islands on board the USS *Arizona* last winter, and I remember that on a certain occasion when

> contact between the battleship and one of our cruisers failed, you very kindly offered before various and sundry representatives of that august body known as the "press" to furnish and afford the Navy an Army plane to assist us in effecting a contact.
>
> In view of your past consideration and helpful spirit, I wish at this time to reciprocate in like manner. It would give me and the Navy Department the very greatest pleasure to place at the disposal of the Army Air Corps a few of the naval patrol flying boats of the scouting force, which are at present temporarily based at Hampton Roads, for your brother service has viewed with sincere appreciation the difficulties experienced by the Army pilots in flying out of sight of land to discover and bomb the old hulk, the *Mt. Shasta.*
>
> These difficulties are the better appreciated since the accomplishment of such type activity has been the common practice for many years of the Naval Air Service. Our material has been constructed and our personnel trained with that end in view. Therefore, the Naval Aviation Service will be glad to either guide or convoy the Army bombers to and from the target, or, if necessary, even undertake the entire mission of finding and destroying by bombs said old hulk.

This letter was just what the press needed to have more headlines.

Naturally, Billy Mitchell was asked to comment and he did—at length. He blamed inadequate equipment and lack of practice and used this to further his argument for a unified air force. Humorous editorials and cartoons appeared. Pro-Navy reporters had a heyday digging up the old clichés. But worse was yet to come. Jim Fechet was furious at the adverse publicity, but we were now fully committed to the test and we had to go through with it.

There was better weather on the fourteenth, and the bombers took off again. This time they found the *Mt. Shasta* and began their attacks at noon. Forty-nine bombs were dropped from different altitudes and in different formations. Only one bomb made a direct hit on the old tub, although most could be classed as "near misses." Dargue, seeing that the *Mt. Shasta* was still afloat, radioed the Coast Guard to sink it to prevent it from becoming a danger on the shipping lanes.

The Coast Guard crew first went aboard and made notes of the damage, then opened the sea cocks. The cutter then fired one-pound shells at the water line to hasten the job, and the old freighter finally sank to its grave late that afternoon. The skipper of the *Mascoutin,* Comdr. Eugene Kiernan, made a full report of the damage and noted that although the bomb that had made the direct hit was not powerful enough to put the vessel out of commission, there were enough holes in the side of the vessel from the bombs dropped nearby to cause its eventual sinking within twelve hours. Dargue also noted, when he forwarded this report, that "it was most unfortunate that no 600-pound bombs were available" and that the results obtained "indicate a disappointing degree of progress since 1921." He pleaded for improved bomb sights and bomb racks, and indicated there had been notable lack of success owing to "lack of sufficiently improved equipment to permit accurate high-altitude and cross-wind bombing."

From this poor showing more lessons were learned, and, as usual, progress came about because of the failures and not the successes. But the press went overboard and without reference to the facts had more headlines:

> ARMY BOMBERS FAIL TO SINK MT. SHASTA
>
> ARMY AND NAVY RENEW OLD FIGHT OVER SHIPS VS. BOMBING PLANES
>
> OLD FREIGHTER, ARMY BOMB TARGET, RESTS ON BOTTOM BUT COAST GUARD PUT IT THERE

The articles were badly distorted and not one newspaper bothered to print the Coast Guard skipper's appraisal of the damage or the fact that the ship would have sunk by itself if left alone. A few days later editorials appeared and used the inaccuracies of the news stories as the basis for their observations. The following excerpt from the *Boston Traveler* was typical:

> It is unimportant that a war game failed, but it is important that with full information about the location of an "enemy raider" the whole system collapsed. Is this what we may expect in case of war? Is this the type of defense we have? Those responsible should be called on the carpet. If we must have wars, let us have them competently conducted.

Jim Fechet had to bear the brunt of this episode, but I did the best I could to take the load off his back. An investigation was conducted, and when the report got to me for reply to the Army Chief of Staff, I decided we should take the knocks we deserved but at the same time make a strong case for ourselves. I admitted that the test should have been canceled when it was found that the bombing could not be carried out in shallow water where it would have been under the complete control of the commanding officer of Langley Field, and when it was discovered that no 600-pound bombs were available. I then blamed the failure of the planes to locate the ship on the first day to inadequate all-weather training and proper instruments, and the failure to sink it on the second try to inadequate bomb sights and bombing practice. I closed my analysis with a plea for the procurement of up-to-date equipment and funds to conduct proper combat crew training.

The *Mt. Shasta* incident died down but left a lingering distrust in the public mind; it had a deeper meaning and an aftermath that has never been divulged before. The Navy publicists took the opportunity to question the needs of the Air Corps for planes, and brought their guns to bear on the next round of hearings for military appropriations. President Hoover had asked the Secretaries of War and Navy to look into "the subject of national defense in directions which might improve efficiency and at the same time bring about imperative economies in Federal service." He noted that "our Army has been reduced greatly and the principles of limitation of naval armament have been extended to cover all types of naval craft.

Yet there seems considerable tendency to further expand the air components of the two services and perhaps regard them as exempt from the application of these principles."

Adm. W. V. Pratt, chief of Naval Operations, had his representatives on the Joint Army Navy Board deliberately drag their feet and question the scope of the enquiry that had thus developed from the President's letter. As a result of the deadlock that developed, General MacArthur gave me the job of making an independent study showing both Army and Navy needs for aviation and to be as objective as possible in my assessments. I made a thorough review of the nation's need for air power and pointed out the basic differences between the needs of the Navy and the Army. I then launched into the probable threats to American security and noted that "unfortunately, it is quite possible for a war to break out which would inevitably involve us." I believed that a war might break out in one of two ways: a conflict initially European into which the United States would be drawn in spite of its efforts to remain neutral or by a declaration of war against us by a coalition of nations. I also noted that "a state of war in an economic sense exists today between a so-called communistic system, as exemplified by Russia, and a so-called capitalistic system, as exemplified by the other principal powers of Europe. There are decided tendencies and distinct dangers of defection of certain important powers from the capitalistic to the communistic side. The Communists will never be able to impose their system upon all the other nations of Europe without a resort to arms."

Among other things, I recommended that the Marine Corps and other land-based combat aviation be eliminated and that the Navy be required to justify the great excess of naval aviation over that of foreign powers, their excessive proportion of observation aviation, and the excessive cost of their pilot training system. As can be well imagined, these were bold words and not calculated to meet the approval of the Navy. My predictions of the possibility of war and the remarks about communism were judged so farfetched that the study was promptly stamped "Secret" and filed away, never to be used.

That report was dated October 9, 1931. My only regret is that I couldn't convince anyone to believe it. However, there were other ways to make the point and I was determined to take advantage of them.

In the summer of 1931 it was widely rumored that I would become the next chief of the Air Corps in December when Jim Fechet retired. Fechet went on a three-month leave in September, and I took over his duties in his absence. On December 21 I assumed command, and received my second star as a major general.

The press and the Congress were good to me when the nomination was announced, As far as I know, no one on the Hill openly objected to me getting the job and the rank that went with it. I will probably never know for sure, but I think the 1931 maneuvers which had come off without a hitch had a lot to do with it. If there had been any mid-air collisions or bad crashes, my career would have ended right there.

When I pinned on my second star, I realized that the high-school dropout who had ridden away from home on a bicycle in 1898 to join the service had come a long way. I had weathered the Spanish-American, Philippine, Cuban, and Mexican conflicts, and World War I. They were nothing compared to the war that lay ahead—on Capitol Hill.

Chapter Thirteen

The Poverty Days of the Air Service

No American in uniform I have ever known disputes the fact that there is, and should be, complete control of our military establishment by civilians. It is one of the principles upon which this nation was founded and has survived. The marvelous system of checks and balances that exists prohibits a military clique from taking over and running the country. Thousands of men have died to preserve that system and will continue to do so when their nation's safety is threatened.

There are times, however, when the military man's loyalty, integrity, and temper are tested to the utmost. One of these occurs when he assumes a position that occasions him to be called to the nation's capital and appear before a committee of the representatives of the people. The nation's duly-elected representatives, unfortunately, are not always well-informed on all matters they are called upon to judge. A notable few seem to take great delight in considering the military witnesses before them as on trial for some dastardly crime. These confrontations are labeled "fact-finding investigations" which, too often, are conducted as kangaroo courts where the witness becomes the accused who has already been adjudged guilty before his "trial."

Before I became chief of the Air Corps, I had testified before congressional committees approximately seventy-five times.

No sooner was I in the chair as the new boss of the Air Corps than I received notice to appear before the House Committee on Expenditures in the Executive Departments on February 4, 1932, concerning the question of consolidating the Army and Navy into one Department of National Defense. Two separate bills had been proposed, both of which provided for the appointment of one Cabinet officer and three assistants—one each for Army, Navy, and Air.

At first it was my intention to appear and merely answer any questions put to me. However, a week before I was to testify a member of the Congress, Representative Charles Martin of Washington and a former assistant Chief of Staff of the Army, had made a number of statements before the committee which were so derogatory to the Air Corps that I could not let them go unchallenged. He criticized us for unwarranted extravagance and said that this extravagance would only proliferate if the air services of the Army and Navy were made an independent organization. "You talk about an independent air force," he had told a reporter. "I have the privilege, as a member of this committee, to speak a little bit freely about that: I tell you that if you turn those air birds loose, you will have to organize something like the Veterans Bureau to take care of the appropriations to keep them going. They are the most extravagant, undisciplined people on earth. They go up in a plane and fly three or four times, and then they go out yelling about flying flaming coffins. If we had followed Mitchell's program, we would bankrupt the government. Those fellows have no sense of economy."

I told the committee that the remarks made by the general indicated "a pitiful lack of knowledge regarding the administration and operation of the Army Air Corps during the past five years, and the nature and tone of these statements call for a definite reply to such unwarranted public criticism of a body of men who, in my opinion, are not exceeded in efficiency today by any other branch of the United States army." I reviewed the impact of the airplane on modern warfare and then summed up my position with regard to a separate Department of National Defense by saying that I could not concur in the enactment of any legislation for such a complete reorganization until a thorough and exhaustive investigation had been made of both the Navy and Army air-expansion programs then under way. My primary reason for such a stand was that I could not see how we could create a Department of National Defense without first having ascertained the relative roles of our land, sea, and air forces. The bills then under consideration did not do this. Since the Air Corps could not even get appropriations for its five-year program because of the constant demands of the Navy air service, I did not see how creating a separate Air Force and superimposing another bureaucracy on top of it could solve things.

The committee questioned me for about three hours on my opinions. One of the significant exchanges took place as follows:

MR. WILLIAMSON: Will the fact that we created a separate Air Force in any way militate against centralizing the command in time of battle or in defending a great frontier like the Pacific coast?

GENERAL FOULOIS: Not, in my opinion, if you assume that the ultimate responsibility for land defense, the defense of our country, upon which we ordinarily predicate our policy, is placed in the command of the man who controls the land forces. We have not got to the point in air war yet where we can start a war and finish it alone; it is some time off; and I concur with General Martin in connection with his statement that there should be just one man running a show in war over certain areas within the limits of which he can control. And under normal conditions that man should be the man who has the land forces under his control.

MR. WILLIAMSON: You would not make it a matter of legislation?

GENERAL FOULOIS: I would not hesitate a minute if I was in charge of the air forces, for instance, in an invasion of the Eastern coast of the United States, if the President were to place one of the Army nurses in charge of those land forces. I would assume it was my job to get under her command and do as she told me under the principle of "unity of command." There should be no two or three independent commanders in one theater of operations where all are working to a common end.

The questions asked by the committee were generally broad in scope, but occasionally would narrow down to specifics about alleged differences between Navy and Army pilot training. For example, there were some Navy air advocates who had said it would be impossible to cross-train pilots of the two services. My views were solicited as follows:

> THE CHAIRMAN: It is your opinion that the Army aviator, if it becomes necessary, should be able to replace the naval aviator, and the naval aviator, if it ever becomes necessary, should be able to replace the Army aviator. In other words, they should be trained so that they could perform duties on land or sea if it was necessary.
>
> GENERAL FOULOIS: I see no reason why a policy like that should not be considered sound.
>
> THE CHAIRMAN: Considerable was said here in reference to navigation. It was pointed out by the Secretary of the Navy that navigation at sea and navigation over land are vastly different. The answers to that could be that men who flew around the world had navigated their ship properly over the land and over the sea; so did Colonel Lindbergh whose flying prior to his trip across the ocean was as an air-mail pilot. This is an indication that it is not vastly different.
>
> GENERAL FOULOIS: I see no reason why land-trained fliers cannot navigate over the sea just as well as sea-trained fliers navigate over land. It is simply a matter of training.

Although I had added a few new detractors to my list by my testimony, I also had some support I was not aware I had. On March 5, 1932, I received a letter from Representative John J. McSwain, chairman of the House Committee on Military Affairs congratulating me on my testimony and added: “It is very gratifying to my feelings that an enlisted man has risen by force of merit to the rank of major general and the headship of the most important branch of our national defense organization. The contrast between the dignity and force of your statement and the confusion and lack of dignity of the statement of a certain blue blood of the Army is very striking.”

I survived this appearance on the Hill and went about the normal business of being head of the Air Corps.

The year 1932 was an election year and Franklin D. Roosevelt, the former Secretary of the Navy with whom I had crossed swords on Capitol Hill in 1919, was swept into the highest office in the land. The country was in the grips of a mighty economic depression, but in spite of this the Air Corps had introduced two new observation planes, two light and two heavy bombers and a new attack plane into the inventory that year. Capt. Albert F. Hegenberger made the first blind solo flight (without a check pilot aboard) solely on instruments, and the incomparable Jimmy Doolittle, a Reserve major at this time, set a new world speed record for land planes by averaging 294 miles per hour over a three-kilometer course at Cleveland. Outside the service some little-noticed developments took place that were to have their effects in the years to come. A scientist named Robert Goddard made the first flight of a rocket with gyroscopically controlled vanes for automatically stabilized flight near Roswell, New Mexico. An altitude record of 43,976 feet was set for land planes by an Englishman. The first teletype printer weather map service was inaugurated and a control mechanism for variable pitch propellers was developed. So, aviation was progressing and the speed, altitude, and distance frontiers were being pushed farther back.

While the hardware was necessary for aviation progress, I realized that it took more than that to win wars in the air. There had to be some master scheme, some grand strategy to it all, some doctrine that would guide the development of the Air Corps in the years to come. I gave the task of shaping the Air Corps of the future to Oscar Westover and the Air Corps Tactical School located at Maxwell Field, Alabama.

Westover, assisted by a young captain named George C. Kenney, sponsored the translation of a book entitled *Command of the Air* written by Lieut. Col. Giulio Douhet, an Italian artilleryman.

Douhet did not agree that wars must be waged exclusively

between the armed forces of nations. He believed, as I did, that Germany's armed forces, especially its air force, had not been defeated in World War I. It was the desire of the German people to cease fighting which had ended the war, and this national will was made up of many factors—economic, psychological, social, and political—as well as military. But Douhet's most unique contribution was his concept in regard to air power. He theorized that air power could influence all these factors because it could be brought to bear against armies, navies, cities, and industrial complexes. He recognized that the principle of surprise was applicable to technical advances. "The form of any war," he wrote, "depends upon the technical means of war available."*

The Douhet theories, along with those of Gen. Hugh Trenchard, commander of the Royal Flying Corps during World War I, were studied at great length by the students at the Air Corps Tactical School and a guiding air doctrine slowly emerged. The seeds for the future were sown there at Maxwell Field in those bleak months of the early thirties. To make these seeds grow would take public support which had to be translated into appropriations by the Congress. The onslaught of the depression, however, put the nation in no mood to increase the military budget. Postwar apathy and the continual demands for disarmament added to the intensified reductions so that by the fall of 1932 the Armed Forces of this country had reached a low-water mark of effectiveness. General MacArthur, our Chief of Staff, then estimated that the United States stood seventeenth in rank among the world's armies. Publicly, I had to agree that the Air Corps was just as bad off; privately, I knew we were in worse shape than that.

None of us knew how the Roosevelt administration would feel about the Air Corps, but I was not very hopeful that a former Secretary of the Navy who had spent much of his leisure time on the water and had never been up in an airplane would be sympathetic to our requests for technical improvements. I passed the word to all units confidentially to continue their efforts to push

* Giulio Douhet, *The Command of the Air,* Coward-McCann, Inc., New York, 1942, p. 6.

the state of the flying art at every opportunity. When we didn't have the funds I made clandestine trips to see the heads of major airframe, engine, and component firms and asked them to keep their engineering talent working and to try to find the money somehow to continue their research. Most of them did, and such men as Larry Bell, Donald Douglas, Glenn Martin, Bill Boeing, and "Dutch" Kindleberger managed to do so, in spite of the fact that they all faced financial ruin in an industry that seemed to be going nowhere because of lack of government support. Those who had the ingenuity and had engendered the deepest loyalty in their employees eventually prospered. They deserved to.

The winter of 1932-1933 was marked by something more than transition from one administration to another in Washington. Unprecedented blizzards swept across New Mexico, southwest Colorado, southern Utah, and northern Arizona. More than 21,000 Indians in the villages throughout the area were isolated and faced starvation. There was no way to relieve their plight-except by air. I fired off a confidential message to "Hap" Arnold at March Field. Could he help out somehow?

The answer was typical of all of my top men in those days. Hap wired back in the affirmative that he would despatch men and planes to drop supplies although none of his men had ever done anything like it before. It was a chance to prove a point for air power and they didn't muff it.

The highly successful mercy missions by Arnold's bombers received favorable publicity but there was nothing we could do to change the grim specter of hundreds of hungry people stalking the streets of the rest of the country simply because they had no money, no jobs, and, apparently, no future. The new administration fell heir to acute economic distress across the country as well as a depleted military establishment. Faced with this, General McArthur asked us all to plan realistically for our new fiscal-year budgets. Translated, this meant we should cut to the bone.

We did not expect the new President to go along with any expansion plans. We were right. Instead, he urged the Congress to cut the War Department budget by $144 million and reduce our personnel strength by retiring 4,000 officers and dropping 12,000 enlisted men.

There was much public support for these cuts, not only from the pacifists and various church organizations, but many newspapers as well. However, there were voices on our side and they were strong.

The Hearst chain of papers supported the program that we wanted and an editorial in the *New York American* summarized the military views by saying that "to destroy what is left of our regular Army, which is the backbone of our defense on land, would be to mortgage the defense and independence of the nation and let a vital part of our national insurance lapse."

While the wrestling for appropriations was going on in Washington, I ordered an extensive exercise to take place in May on the West coast. In January 1933, the War Department had finally given the Air Corps a clear statement that its mission would include long-range reconnaissance and operations "to the limit of the radius of action of the airplanes." The purpose of the exercises, then, was to show that we could concentrate our air units on the coasts quickly. We hoped also to prove the need for aircraft capable of long-range operations against enemy fleets approaching our shores. Up to this time my air staff planners had been stressing the need for bombers in offensive operations and had gotten nowhere with the General Staff. Discouraged, they came to me and I could see only one way out: "Stress *defense,* not *offense,* and stress re-enforcement of the Hawaiian Islands," I told them, "and maybe that will work." It did, and the climate became more favorable as we dropped the word "offense" from our justification papers. Thus, historians will note that our plans continually stressed the *defense* of our coasts and our overseas possessions from that point in time on. As I saw it, if we could get bombers that could carry bigger bomb loads and fly greater distances this way, what difference did it make what words we used?

It was this slow accretion of time, circumstances, and rationalizing that finally sold the idea of forming a General Headquarters (GHQ) air force that would provide a central striking force under the top command of the Army. Such a force could not operate to defend our coasts and overseas possessions without bombardment and observation planes.

Naturally, it took a lot of doing to get the Navy to go along with the idea of the Army operating off the coasts at sea. General MacArthur, however, had worked out a compromise with Adm. William V. Pratt, chief of Naval Operations, in December, 1931. They agreed that naval air forces "will be based on the fleet and move with it as an important element in performing the essential missions of the forces afloat." The Army land-based air arm would defend the coasts and overseas possessions, thus leaving the fleet assured of absolute freedom of action with no responsibility for coastal defense. This dividing line enabled each service to proceed with its own planning, training, and procurement with little possibility of duplicating the activities of each other. Such an agreement hadn't been easy to obtain, and MacArthur noted in his annual report that "at times in the past the difficulties blocking the path leading to a practicable agreement in this matter seemed almost insurmountable."*

On January 18, 1933, the GHQ Air Force (Provisional) was organized and placed under the command of Oscar Westover. It was basically to test the value of such an organization under wartime conditions that I authorized the West coast maneuvers.

I gave Westover a completely free hand, and his report was one of the most valuable documents on the employment of air power up to that time from the standpoint of lessons learned and constructive criticisms. He concluded that all Air Force units could be concentrated on either coast within two and one-half days for the nation's defense and stressed the need for "adequate and suitable equipment and material" for over-water reconnaissance. More importantly, he made excellent suggestions for the organization and command of a GHQ Air Force as a permanent part of the national defense.

Probably the most significant observation from a historical

* Annual report of the Chief of Staff in *Report of the Secretary of War to the President,* 1931, p. 38.

viewpoint that Westover made was to regard reconnaissance and bombardment as the primary functions of a GHQ Air Force, noting that "high speed and otherwise high performing bombardment aircraft, together with observation aviation of superior speed and range and communications characteristics, will suffice for the adequate defense of this country." He explained the reason for this observation:

> During these exercises, observation aviation appeared woefully obsolete in performance, as did pursuit aviation in speed characteristics. Since new bombardment aircraft possesses speed above 200 miles per hour, any intercepting or supporting aircraft must possess greater speed characteristics if they are to perform their missions. In the case of pursuit aviation, this increase of speed must be so great as to make it doubtful whether pursuit aircraft can be efficiently or safely operated either individually or in mass.

This statement, naive in retrospect, is significant in the light of later events because it shows the basis for the conflict between the bomber boys and the pursuit enthusiasts. He found that the B-l0s could out-speed the P-12s, which caused Westover to rationalize that "bombardment aviation has defensive firepower of such quantity and effectiveness as to warrant the belief that with its modern speeds it may be capable of effectively accomplishing its assigned mission without support." He added that "the ability of bombardment aviation to fly in close formation and thus to ensure greater defense against air attack, together with improved efficiency of silencers and camouflage, allows no known agency to frustrate the accomplishment of a bombardment mission."

So it was this rationale, plus our clearly-defined mission giving us the right to engage in seaward reconnaissance that enabled the Air Corps to begin development of long-range bombers in 1933. Although our 20/20 hindsight shows that we should not have done so at the expense of pursuit aviation, I have no regrets today. My decision then enabled us to have B-l7s and B-24s by the time we entered the war in 1941 and provided the basis for the B-29s that ended it in 1945.

The limited funds available caused us to look at bomber designs very carefully. In July 1933, I received an engineering study from Wright Field directed to the question of design for maximum range with a 2,000-pound bomb load. The study indicated that it was not beyond the state of the art to build a bomber that could combine the functions of reconnaissance and bombardment in one air frame and build into it a range of 5,000 miles at 200 miles per hour. Accordingly, in December of that year I signed and forwarded "Project A" (XB-15) to the War Department, which was the go-ahead for such a plane.* The support for this proposal was summed up in my statement that it would "not alone reinforce either coast line, but would definitely enable reinforcement of Panama and Hawaii."

Much to my surprise, seven days after I forwarded the proposal, the General Staff authorized tentative approval of $609,300 for long-range bomber development. The following February the development was approved "in principle" and in May 1934, the Chief of Staff authorized the negotiation of contracts with the Boeing and Martin companies for preliminary designs. A few days later, at my request, the military characteristics were adjusted to a tactical mission for "the destruction by bombs of distant land or naval targets." I also added the language that the purpose of the bomber was "to reinforce Hawaii, Panama, and Alaska without the use of intermediate servicing facilities."

By June 1934, Boeing had completed its design study, and a year later a contract was signed with Boeing for the purchase of one XB-15, which we thought would be our first operational four-engine bomber.** Although it did not fly until 1937, flight tests showed that the design engineers were too ambitious and

* Projects B, C, and D later became the B-17, P-38, and B-19. The latter was the forerunner of the B-29.

** The Witteman-Lewis NBL-1 (Barling) bomber actually had six engines in four nacelles. Designed in the early twenties, it had disappointing speed, load, and endurance characteristics.

optimistic. They had planned for larger engines to carry the size and weight they envisioned, but the engine developments required were not forthcoming. However, both our engineers and Boeing's learned a valuable lesson. They had tried to push the state of the art too far and too fast. There had to be an interim step to planes as large and heavy as the XB-15.

While the XB-15 contract was being negotiated, we distributed design proposals among all the plane manufacturers for a multi-engine bomber asking them to compete for a contract the following year. Most of those who wanted to compete made the assumption that we wanted a twin-engine plane, since we had already committed ourselves to the XB-15. Boeing, however, did not make that assumption. While looking for a transport to compete with the highly successful Douglas DC-2s, Claire Egtvedt, their chief designer, suggested they consider both a transport and a bomber of parallel design at the same time. Both would be midway in size between their B-9 and the XB-15 and have four engines instead of two. Would the bomber version meet the Air Corps specifications?

We had purposely not specified the number of engines because what we wanted was performance. The invitation to bid we sent out was for the following: bomb load, 2,000 pounds; required top speed, 200 miles per hour; *desired* top speed, 250 miles per hour; required range, 1,020 miles; *desired* range, 2,200 miles; crew capacity, four to six. We anticipated the winner might get a contract for 220 planes. We asked that a flyable plane be presented for testing by August, 1935.

Two other manufacturers decided to compete—Martin and Douglas. Both of them put their engineers to work on twin-engine models, the latter on a bomber version of the DC-2. My people at Wright Field watched all three of these planes with great interest, and when the summer of 1935 rolled around, everyone on my staff was excited about the new hardware we might soon acquire—the Congress willing.

The newspapers caught our feeling of excitement and interest centered on the four-engined Boeing which we had designated the XB-17 (Project B). It was shown to the press in Seattle on July 28, 1935. A silver giant bristling with five machine-gun turrets, it was immediately dubbed “a veritable flying fortress,” and the name stuck.

After preliminary flight tests it was flown nonstop to Wright Field, a distance of 2,100 miles, at an average speed of 232 miles per hour. When this news reached my office, I felt an inner satisfaction I had never known before. American free enterprise had accomplished what many had said was impossible. Just as many had said the Wright brothers could never meet the specifications we had set for their *Flyer* in 1908, there were doubting Thomases who predicted that the XB-17 was too heavy to fly and even if it could fly, it could never fly at 200 or more miles per hour.

Just as tragedy marred the first test of the Wrights at Fort Myer, the XB-17 had its rendezvous with fate. The tests began on schedule in the summer of 1935 with the Martin B-12, the Douglas B-18, and the XB-17 in competition. Each went through the evaluations during the next several weeks according to our rules: speed, endurance, time to climb, service ceiling, armament, maintenance, flying characteristics, and installation of equipment. The tests were uneventful until on October 30 the XB-17, piloted by Maj. Ployer P. “Tex” Hill, took off, went into a steep climb, stalled, and crashed. Hill was killed, as was the Boeing observer on the flight, Les Tower.

The crash made national headlines, and there were many articles with an “I-told-you-so” twist to them. “It was too much airplane for one human being to handle,” they said. The truth, which never seemed to get printed, was that the ship had taken off with its elevator control surfaces locked. Control surface locks were new because Boeing had found that the movable controls were so big on the XB-17 that they could be damaged on the ground in high winds and also made the plane difficult to taxi. Their engineers had designed a lock that could be applied inside the cockpit. Apparently, the crew had not checked this item before takeoff—a fatal human oversight which has killed other men in other planes ever since.

I had followed the reports of the tests closely and was determined that human error was not going to doom the plane that looked like the answer to our dreams. I approved the purchase of the Douglas B-18 because it had passed all the tests well. But I also approved an order for thirteen B-17s plus a fourteenth which we would ultimately submit to extensive destructive structural tests. I had no way of knowing then that those thirteen B-17s would be the only four-engine bombers in the hands of the Air Force on the day that Hermann Goering ordered his Stukas against Poland in September, 1939.

The budget battle on Capitol Hill in the spring of 1933 was a bitter one. The new administration was determined to cut expenditures for defense on the one hand and dedicated to the beginning of a massive giveaway program on the other. They called it a New Deal, but as far as I was concerned it was the same Old Deal where the rules of the game were such that the Air Corps was to continue to be the step-child of the Army. One of the first assaults in the new economy wave was made against flight pay for military aviators.

As chief of the Air Corps, I had to prepare a defense of the flight pay scale then in effect and fight those who thought that the flight pay system was illogical.

It was not difficult to find evidence of the need for extra compensation. We discovered that while the Air Corps personnel strength was only 10 per cent of the total Army strength, we had 37.7 percent of the peacetime fatalities. Our accident rate was triple the expected average fatalities among men between the ages of twenty-five and fifty which would make an extra annual insurance premium of more than $30,000 necessary for the average pilot. Because of the necessity for us to make regular and frequent flights in military aircraft, additional premium for flight hazard was required by the insurance companies—provided they would grant the insurance at all. We figured the average pilot would have to pay from $850 to $1,500 annual premiums, depending on his age, to protect his family adequately. If the flight pay were dropped, he would still have to fly and he would still feel the obligation to buy insurance. No matter how we figured it, about four out of every hundred of our pilots would be killed each year. If flight pay were cut so that pilots could not be compensated adequately for their extra risks, I knew we would have a king-sized morale problem on our hands. Without properly motivated crew members, the acquisition of the finest planes in the world would not make an air force.

As the only two-star general drawing flight pay, I came under personal attack for receiving annual flight pay amounting to $4,000. The law allowed it, and I took the risks to earn it. In fact, during that year I flew 234 hours, most of it solo. Only 40 pilots could top that figure and most of them were captains. No one above the grade of major could match it, and only three majors flew more than the "Old Man" that year.* Once these statistics were furnished, criticism against me immediately ceased and again focused on the main issue.

While the requirement to justify flight pay was bothersome and time consuming, it gave us a chance to explain to ourselves why we needed the extra pay and I think that some good came of it all. The advent of safer transports, such as the DC-2, made civilian flying less risky and the accident rate for commercial planes was better than that for military flying. But the commercial flier operated over fixed routes, over established airways dotted with emergency landing fields at frequent intervals, guided by lights and marked airways, and kept informed of weather conditions by radio.

The military flier had none of these advantages. He flew mostly over uncharted courses in adverse weather. He carried high explosives. He had to practice unusual and dangerous maneuvers and subject his plane to great stresses at high speeds. He had to control his plane not only when flying alone but also flying in close formation. He had to fly dangerously close to the ground and at such high altitudes that he could not survive more than a few

* Majs. Gerald E. Brower, William E. Kepner, and Clarence L. Tinker.

seconds without oxygen equipment. In short, it was completely unfair to compare the civilian pilot's risks with those of the military pilot. There has been no change. Military flying is still more hazardous today than any kind of civilian flying and the extra pay is still much deserved.

The pressure on the services to come up with budget reductions was too great. Flight pay and allowances were reduced by about 10 percent Many officers were furloughed and we did not have the funds to pay new second lieutenants who had just completed flying school. Our requests for funds for new planes were also drastically cut. The pinch was on. The military budget for fiscal year 1934 was the expression of a people completely disinterested in a war in China and the assumption of power in Europe by men named Mussolini and Hitler. The feeling was that we could still hide behind our oceans. We already had planes and pilots. The Air Corps could wait until times were better. Therefore, it was disloyal and unpatriotic for military men to stand before the Congress and argue for funds to prepare for a war that could not possibly happen when the nation was experiencing the worst depression in history. We were more than 1,000 planes short for minimum defense needs and those we did have were quickly becoming obsolete. We needed more than 6,500 officers and men. We were not to get them.

The word was quietly passed from the White House downward through military channels. Cease resisting the inevitable and live with what you have, we were told. We were to do nothing to attempt to arouse public sympathy for the ever-dwindling potency of the military services.

The fall of 1933 turned into a bitter winter—the bitterest on record. It was certainly to be the bitterest winter in the history of the United States Air Force.

Chapter Fourteen

The Truth about the Air-mail *'Fiasco'*

There are three dates that I think historians should note as the most significant dates in the development of American air power. The first of those is December 17, 1903, when the Wrights proved flight in heavier-than-air machines was possible. The second is March 19, 1916, when the 1st Aero Squadron took off for Mexico on the first air combat mission in United States history. The third date is February 9, 1934, a day not marked by any flying feat but a three-hour conference I had with the Second Assistant Postmaster General Harllee Branch, who had been appointed to the job only three weeks before.

The background to that conference went back to 1918 when the Air Service had inaugurated air-mail service between New York and Washington. Between May and August that year a half-dozen Army pilots flying JN-4s had proved that it was possible to carry mail safely and on a regular schedule. From that time until 1927 the Post Office extended and operated air-mail routes from coast to coast and border to border.

In 1925 the Congress passed a bill called the Kelly Act "to encourage commercial aviation and to authorize the Postmaster General to contract for air-mail service." This legislation changed the concept of the government maintaining the planes and paying the pilots and put the mail carrying in the hands of privately owned and operated companies. The result was the beginning of the scheduled airline business because it provided the necessary incentive for aviation interests to form operating companies and invest in their growth.

In a very short time a number of contract air-mail lines were operating over hundreds of miles of air-mail routes. These lines at first flew only the mail because the revenue was sure and passengers were a nuisance. In fact, few air-mail planes had any extra seats for passengers at first. But under government prodding and with the realization that passengers would pay well for the privilege of riding with the mail, more and more passengers began to be carried. Later, the Congress changed the basis of payment from weight alone to the volume of space aboard a plane available for mail. These factors encouraged the contractors to buy larger planes that could accommodate both passengers and mail.

As air-mail contracts expired during the Hoover administration, Postmaster Gen. Walter F. Brown consistently favored those lines with the strongest financial backing for renewals. The smaller lines not so favored had to merge with larger lines or go out of business. The low bidders did not always get the contracts and Brown ignored the pleas of independent operators for a chance to prove themselves. He was accused of having needlessly given the big operators more than $47 million which represented excess payments over the low bids and the contract prices granted.

The Roosevelt administration was embroiled in the uproar created by the independent operators as soon as it was in office. A congressional investigation discovered some disquieting facts in regard to air-mail contracts which indicated a "wholesale conspiracy to defraud the government." In January 1934, Solicitor General Karl Crowley, investigating the many allegations for the Post Office Department, made a preliminary report that was leading to a recommendation that all air-mail contracts be canceled and new ones be let "to the lowest responsible bidders carrying the mails to the end that there may be no interruption in air-mail service to the people of the United States."

On February 8 Postmaster Gen. James A. Farley, accompanied by three assistants, met with President Roosevelt to apprise him of the evidence they had found. What the President heard then, plus what the Senate investigating committee had dug up, must have made him furious. Ever since he had come into office he had crusaded against governmental corruption and favoritism in the granting of contracts. He was determined to act, recognizing that

1934 was an election year and that the stew that had been concocted smelled unrosy. Cancellation could be effected with a stroke of the pen. But who would then fly the mail until new contracts were negotiated?

I had been watching the air-line scandal in the newspapers but assumed, naively, that when the mail couldn't go by air, it would go by rail until things were straightened out. Mail had gone by rail before airplanes started flying it in 1918. It would still get through. A few businessmen might suffer, but the economy of the country wouldn't collapse so I wasn't particularly concerned when the air-mail contract pot began to bubble on Capitol Hill.

About eleven o'clock in the morning of February 9 I received a telephone call from Harllee Branch. Could I drop over to his office after lunch to discuss some aviation matters? I said I could. After all, I had recently been appointed to serve with Branch and Eugene Vidal of the Department of Commerce on an interdepartmental committee on aviation policy. I assumed that he was calling a meeting to go over some of his ideas about the national airways we were hoping to improve or future legislation we wanted to propose.

When I arrived at Branch's office he said he wanted to chat with me about air-mail runs, schedules, and some internal-management problems they had encountered. He mentioned the matter of the cancellation of the air-mail contracts and casually asked, "If the President should cancel all contracts, do you think the Air Corps could carry the mail to keep the system operating?"

I thought a moment before answering. We were in sad shape in the way of cargo equipment and none of our pilots had flown regular schedules on instruments or at night. However, we might use our bombers and observation ships and our pilots would get some badly needed training which had been curtailed because of the shortage of funds to buy gasoline, equipment, and spare parts. Besides, if the President wanted us to do the job, he was our Commander in Chief. The problems we might have would focus national attention on the Air Corps and maybe we would then get the funds we needed for expansion....

"Yes, sir," I answered. "If you want us to carry the mail, we'll do it."

Branch seemed relieved and for the next three hours we discussed the various route schedules in detail. From what I heard, I did not see any great difficulties. We had enough planes provided our pursuit planes could be put on the short runs with the light loads. We would not be expected to operate the entire system then in effect nor maintain the regularity of the airline schedules. Neither would we be expected to handle the volume and poundage that the private carriers were flying.

Since many of the airline pilots held reserve commissions, I suggested that the President could order them to active duty. This would give us a nucleus of experience upon which we could build because these men could be instructors and give us the benefit of their thousands of hours of airline flying. I assumed that all of the airport and airways facilities throughout the country would certainly be put at our disposal and that the necessary funds would be made immediately available. With all of these factors assured, I considered the possibility of temporary mail duty as the ideal peacetime test of our men, planes, and organization to meet a national emergency.

"How long do you think it would take you to get the Air Corps ready?" Branch asked.

"I think we could be ready in about a week or ten days," I answered casually, thinking that he certainly didn't mean how much time *from that moment on.*

The discussion had been interesting and stimulating. The obvious next step for me was to get my staff together and give them the task of writing a contingency plan which could be presented to General MacArthur and put into effect if and when the President should go ahead with the contract cancellation.

I returned to my office in the War Department and gave my staff a briefing, assigned them their separate planning tasks, and sent them scurrying. I then left my office to report the meetings with Branch and my staff to Gen. Hugh Drum, deputy Chief of Staff. Just as I was about to enter Drum's office, General MacArthur appeared. His face was flushed and I sensed immediately that I was precisely the man he was looking for and that I was in trouble.

"Foulois," he barked, "a newsman just told me that the President has released an Executive Order giving the Air Corps the job of flying the mail. What do you know about it?"

I was dumbfounded. I had just struck a match to light my pipe the moment before he appeared and his question caught me in mid-flame. His sudden appearance, his stunning question, and the burning match caught me at the same instant. I must have leaped eight feet in the air as my senses of sight, hearing, and touch were so suddenly stimulated. I returned to earth in a shower of sparks and tobacco which also fell on the boss. I tried to explain that I had just returned from a chat with the Second Assistant Postmaster General but that nothing definite had been decided as I brushed the two of us off. I could see I wasn't getting through to MacArthur and I suspect that he had all kinds of black thoughts about my deliberately planting the idea of the Air Corps flying the mails in the President's mind somehow. I tried valiantly to explain that I was as much in the dark as he was.

"Foulois, that is all academic now," he said. "The question is 'can you do it?'"

"Yes, sir, we can," I answered. "That is, if the President will see that we get what we need in the way of support—moral as well as material. It won't be any picnic but we can do it along the lines Branch and I discussed."

"You're on your own now, Foulois. Yell when you need help from me—and keep me informed. It's your ball game."

I was thus dismissed, and returned to my office still in a state of shock. For the next forty-eight hours I held meeting after meeting with my staff and hammered away at the thousands of questions that had to be answered. There was a reason for urgency because Postmaster General Farley had telegraphed a terse announcement to all the private air-mail operators that all domestic air-mail contracts "are hereby annulled, effective midnight February 19, 1934."

It was to be many years before I knew what really had

happened on that fateful afternoon of February 9. While I had been talking to Branch, President Roosevelt was holding a Cabinet meeting at which the subject of the air-mail contracts must surely have come up. In any event, Postmaster General Farley and the Attorney General stayed behind after the Cabinet meeting and a vital decision was reached. A few minutes later the President, taking Farley with him, appeared at his usual afternoon press conference and announced that "the Secretary of War is to place at the disposal of the Postmaster General such airplanes, landing fields, pilots, and other employees of the Army of the United States as are necessary for the transportation of mail during this emergency...."

Apparently all that was needed was a date for the contract cancellation and assumption of the job by the Air Corps. I had unwittingly given that date to Branch when I said I thought we would need "about a week or ten days." Obviously, he had quickly called Farley who, in turn, had told the President, and the date of February 19 was fixed.

The die having thus been cast, it was impossible for me to do anything but carry out the will of my Commander in Chief. The order had been given on a Friday afternoon. The next day General MacArthur gave me special authorization to put direct control of the Air Corps mail operation in my hands so that I wouldn't have to go to him every time I wanted to get something done. The following Monday I knew what units I could call upon and the types of planes we would use. It was up to someone else to tell me what routes we would fly. I met with Eugene Vidal and Harllee Branch and we decided that the priority would be given to the use of air-mail by commercial banks and that such service must be continued to link the twelve cities in which Federal banks were located. Beyond that, all we could tell the press was that the frequency of the flights might be reduced and some routes would be discontinued.

As could be expected, the airline operators screamed loud and long at the cancellation. The press, however, seemed evenly divided over the decision for cancellation. The *Cleveland Plain*

Dealer summed up the "for" side by saying "it is harsh, but the situation it aims to correct is foul." On the other hand, the *Boston Herald* believed the move was "unnecessarily drastic" and the *Baltimore Sun* editorialized that "considerations of a moral as well as a legal nature may dictate a modification of the President's stand."

The air lines recognized that they had to fight to survive. Kept alive by the air-mail subsidy, they knew that passenger traffic alone could not keep them in the black. There was nothing to do but muster their forces and use every means at their disposal to discredit the President's decision and the Air Corps' ability to take up where they had left off.

Two of the most famous names in aviation—Charles A. Lindbergh and Eddie Rickenbacker—happened to be associated with two of the major airlines. Lindbergh immediately captured the headlines when he sent a telegram of protest to the President stating the action taken had condemned the commercial operators without a fair trial. "Your present action," he wrote, "does not discriminate between innocence and guilt and places no premium on honest business."

As soon as the message was despatched to the White House, Lindbergh's lawyer released it to the press. White House Press Secretary Stephen Early, seeing the headlines Lindbergh had made before the President had actually received the message, responded that it was common practice "to allow the President… the courtesy of receiving and reading their communications before they were read by others.... " Early accused Lindbergh of sending the message purely for publicity purposes and reminded him that it was the Postmaster General, not the President, who had canceled the contracts and therefore the letter would be referred to Farley for reply.

Pressure against the decision built up, and even Will Rogers was moved to comment that "it's like finding a crooked railroad president and then stopping all the trains." It was left up to the Postmaster General to defend the move.

As far as I was concerned, the politicians and the

commercial operators could argue all they wanted. I met daily with Branch and Vidal and we slowly worked out details. We agreed that the Post Office Department would transfer $800,000 to the War Department for air-mail operations only to be advised that such a transfer of funds was illegal. Being an optimist about such things, I was not at all concerned that adequate funds would not be made available. Accordingly, I asked the Adjutant General to have all personnel cite the Executive Order in any contracts they might make for the air-mail operation. Little did I realize that my optimism was unfounded and that many of my pilots and mechanics would suffer severe financial hardships as the result.

It was finally decided that we would operate over twelve routes across the country as compared with twenty-six flown by the air lines. There were seventy stops on our routes which extended over 13,284 miles of airways. I established emergency headquarters in Washington and put Oscar Westover in charge.

The entire route structure was divided into three zones and we put Maj. B. Q. Jones, Lieut. Col. Horace M. Hickam, and Lieut. Col. "Hap" Arnold in charge of the eastern, central, and western zones, respectively. Two hundred officers, 324 enlisted men, and 122 planes were identified and immediately alerted. Later I added 50 more men and 26 planes to the call-up. The planes were stripped of armament, seats, and unessential equipment. The rear cockpits of observation planes were converted into mail compartments. Baggage compartments of attack and pursuit planes were modified where possible to make a few more cubic inches of space available. The bomb bays of the bombers were sealed. Radios and blind-flying instruments were hurriedly installed in some of the planes. As soon as planes were ready, they were despatched to the control stops. By the fifteenth pilots were making familiarization flights along their routes.

Although I was satisfied that we would make our deadline of February 19, there were outsiders who were sowing seeds of doubt that we could do the job satisfactorily. One detractor was

Eddie Rickenbacker, who told the press on February 13 that "either they are going to pile up ships all the way across the continent or they are not going to fly the mail on schedule."

As soon as Jones, Hickam, and Arnold got in position at their respective headquarters at Mitchel Field, Chicago, and Salt Lake City, the press descended on them and the quotes were characteristic of the men. Jones, one of the original air-mail pilots when the Army inaugurated air-mail service in 1918, quipped to a reporter that "we'll carry the mail—don't worry about that—unless an elephant drops on us. If it does, we'll cut it up and ship it out as mail."

Horace Hickam, while establishing his office in the National Guard hangar at the municipal airport, was heard to say, "we're not mail-men ... but we've got a job, and we're going to do it." One reporter, surprised that Hickam didn't have a large staff and lots of paper work flowing, noted that his "hat was his office," which was typical of our whole operation since we had neither funds nor logistical support from the rest of the Army.

Arnold, when questioned about whether his pilots could match the airline pilots, refused to concede that the airlines had an edge on flying skill. "Look them over," he said, "you'll find that 90 percent of them were trained in the Army."

But Arnold and the rest of us knew that very few of our pilots had extensive instrument and night-flying experience. The reason was a simple one: no equipment. We did not have the latest instruments and not very many of our planes had landing, navigation, or cockpit lights. The techniques of flying the newly-developed radio beams were devised by the airline pilots and our pilots were not very adept at using them. Without instruments and radios, there wasn't any way for them to practice.

I was concerned that some of the younger pilots would take unnecessary chances in order to get the mail through, so on February 16 I sent a radiogram to the three zone commanders telling them they would "govern their operations with a view to safeguarding lives and property *at all times,* even at the sacrifice of air-mail service. Steps will be taken to inculcate all personnel engaged in air-mail operations with the above policy." I further directed them to consider carefully the "experience of personnel, suitability of aircraft, night-flying equipment, and blind-flying equipment."

Almost as this message was being received, a message was en route to me telling me that three fatalities had just occurred on training flights in Hap Arnold's zone that day. Lieuts. J. D. Grenier and E. D. White had crashed to their deaths in a snowstorm. While on a route familiarization flight in an A-12 between Cheyenne and Salt Lake City, Lieut. J. Y. Eastham, flying alone, had crashed near Jerome, Idaho, in a twin-engine B-7. Weather had been a contributing factor in both crashes. Eastham's accident occurred at night in a fog.

I immediately sent a follow-up radiogram the next day telling the zone commanders to "impress most emphatically upon all pilots the utmost necessity for extreme care and judgment in flying the air-mail. Drill these instructions into your pilots daily until they thoroughly understand the safety-first policy of the War Department and the Air Corps."

At the time of the two fatal accidents, Eddie Rickenbacker was in Los Angeles preparing to fly the last commercial air-mail flight in TWA's new Douglas DC-1 coast to coast. He told the press that the death of Grenier, White, and Eastham was nothing but "legalized murder" and predicted that others would die if the administration persisted in making the Army fly the mail.*

On the morning of February 17 I had a conference with Secretary of War Dern and received his concurrence that there would be no change in my policies just because of the fatal accidents. On the eighteenth I received confirmation from Jones, Hickam, and Arnold that their planes and pilots were all in position and ready to go. That same day, in an obvious effort to furnish a striking comparison with what was to come when the Army took over the mails, Rickenbacker, with two other pilots and a number of newsmen, left Los Angeles for New York. The flight was billed

* I've often wondered whether or not Eddie considered his victories over twenty-six Germans in World War I as "legalized murder."

as the "last load of contract mail" and attracted the attention Rickenbacker wanted. The DC-1, taking advantage of a strong tailwind, was landed at Newark at 10 A.M. on the nineteenth after thirteen hours and four minutes of flight—a new coast-to-coast record.

As luck would have it, Rickenbacker landed just ahead of a snowstorm. By noon snow, rain, and fog moved into the northeast area grounding everything. In the Western Zone a snowstorm slashed through the Rocky Mountain states. Rickenbacker was hailed as a hero in the papers for his "triumphant finale" and we took over the job officially with the worst possible flying weather forecast for the upper half of the country.

The flights in the South and Midwest took off on schedule and completed their first runs without incident. In the West, however, the first pilot who left San Francisco with the Eastbound mail could not get through the pass of the Sierra Nevadas because of the weather and returned to Sacramento. Mail from Los Angeles was put on the train for Salt Lake City.

A few other first flights had difficulties but, surprisingly, none were because of the weather. One pilot, C. R. Springer, took off from Cleveland at night in his P-12 only to return fifteen minutes later. He taxied up to the hangar and shouted to a mechanic, "Get me a flashlight so I can find my way out of this damned town." His directional gyro was inoperative and he could not see his magnetic compass because of poor cockpit lighting.

In spite of these small difficulties, the first day's operation was relatively successful considering the weather. But I was wary of things going too smoothly. True, the men and planes were in position and were getting through, but how long could they continue? I felt that it was inevitable that there would be an accident. When?

The inevitable occurred on February 22. Lieut. D. C. Lowry, flying an Eastern Zone route, crashed south of Toledo, Ohio. An experienced pilot, his death was attributed to a faulty radio since he was 50 miles off course and on instruments. His was the first fatality with mail aboard but not the only accident for that day. Lieut. H. L. Dietz, flying from Washington, became lost in fog and heavy rain at night over Maryland. He circled over the small town of Marion Station until a number of local citizens drove to a pasture and turned on their auto headlights. Dietz lined up with the "runway" but struck the top of a tree on the final approach and lost control. His plane was flipped sideways and struck a telephone pole. Dietz was knocked unconscious but survived the impact.

Just after these two reports reached me in Washington, I received another one from Texas. Lieut. F. I. Patrick had been killed near Denison while en route to supervise pilots on the Midwest runs. He was alone and carried no mail.

In all three accidents weather had been a contributing cause. There was no doubt that the lack of proper instrument and radio equipment and blind-flying experience were also major factors. As could be expected, the newspapers leaped on these accidents as proof that the Air Corps could not do the job.

The next day another accident occurred. A plane ferrying a group of mail pilots was forced down off a New York beach. Lieut. C. F. McDermott, one of the passengers, drowned. Again the papers reacted and referred to the deaths of Patrick and McDermott as if they had occurred on regular mail runs. On the same day, however, a United Air Lines plane crashed in Utah, killing eight people, and little was mentioned of this tragedy in any of the country's papers.

The press reaction to the Air Corps accidents caused a furor on Capitol Hill, and the Republicans and Democrats put on a six-hour debate over the Brunner Bill in the House of Representatives. This bill was to give us the funds and provide the necessary emergency authority to the Air Corps. The Republicans took up Rickenbacker's slogan and accused the administration of "legalized murder." They vigorously condemned the decision to send the Army pilots out with inadequate and unsafe equipment. Representative John Taber announced that he could not in good conscience be a "party to a murder." Several of his colleagues tried to add amendments to the bill that would require every Air Corps plane flying the mail to be equipped with a full set of blind-flying instruments and a two-way radio.

The Democrats, recognizing that the issue had political ramifications in an election year, prevailed. To agree to the amendments would have put the Air Corps out of the air-mail business and might have forced a resumption of the air-line contracts. The bill passed, but I noted that 101 congressmen did not vote.

I was just a pawn in the political chess game, but until I received orders telling me otherwise, I had to keep my men flying. However, to tighten up my safety instructions further, I sent a most urgent radiogram to the three zone commanders on February 24:

> Only pilots of more than two years' service in the Air Corps will be used on air-mail operations involving night flying unless weather conditions all along the route to be flown are excellent. After taking off on a night air-mail run, no pilot will proceed on his flight unless his flight instruments are working satisfactorily and he is receiving proper reception on his radio, Pilots on night runs will not commence flights under unfavorable weather conditions, nor will they continue flights into unfavorable weather conditions.

I further pinned responsibility for plane inspection on the station control officers and required them personally to check each plane that passed through their hands before each flight. I also forbade flights into known icing conditions because we didn't have a single plane equipped with anti-icing or de-icing equipment.

The criticism mounted by the hour. The only friendly voice I heard came from a strange quarter. On February 17 Clarence D. Chamberlain, noted distance pilot, wired Postmaster Farley informing him, perhaps for the first time, that United Air Lines had recently had four crashes in which twenty people had been killed. He pointed out that the press, radio, and newsreels were instilling the belief in the public that the administration made a grave error in canceling the contracts and was doing all it could to "magnify Army mishaps." He was right. But the reporters who were doing the writing had their orders. Since many of the leading newspapers represented Republican interests, their bosses had told them to exploit everything they could find which would prove that the Air Corps couldn't replace the commercial operators. I discovered this when a group of reporters came to my office for an interview after McDermott's death from drowning. The reporters seemed especially hostile that day when they arrived, and I asked them why. I had known many of them since I had come to Washington in 1927, and I had considered most of them pro-Air Corps.

"Well, General Benny," one of them finally told me, "we have orders to 'get' you. We're supposed to dig up all we can to prove you're inefficient and that Army fliers are not equipped to do the job."

"Well, I know what it is to obey orders," I told them. "So go as easy on me as you can. I don't mind if you call me 'inefficient' but remember that the planes are *your* planes and if they're poorly equipped, it's because *you* didn't give us the proper equipment. You might be doing the whole nation a favor if you would slant your stories to show that we should be supported in the Congress when we ask for money to buy new equipment."

I think those reporters did go as easy on me as their bosses would let them, but the flood of mail from concerned citizens and the number of queries from Capitol Hill proved that there was widespread misunderstanding of the facts. Only one man had been killed while actually flying mail up to that time. What could I do to set the record straight?

I visited General MacArthur and told him of my concern. "General, would you give me permission to go directly to the people?" I asked.

"In what way?"

"On radio first, and then give the text of my statement to the press," I answered.

MacArthur smiled and nodded his approval.

"Do you care to see my statement beforehand?" I asked, as I got up to leave.

"Benny, ever since you made that decision not to crash into my tent that day in San Antonio in 1911, I've had a lot of faith in your judgment," he said with a broad grin. "Besides, I told you this was your ballgame!"

I was reassured by this parting comment and made arrangements to broadcast over the CBS network through Station WJSV at Alexandria, Virginia, on the night of February 27. I began by reviewing how the Air Corps had been given the air-mail task and how I had organized my resources to do the job. I then explained that our planes were now being equipped with radio receivers and told how radio beams worked on the airways.

There was no use denying that we had to overcome handicaps but I pointed out with all the hullabaloo about "legalized murder" that only one pilot death had occurred on a scheduled air-mail run. I read excerpts from my "safety-first" radiograms to the zone commanders and then answered those who criticized the ability of our pilots to carry out their flying assignment:

> Much has been said lately of the type of Air Corps officers now actually flying the mail. Some would have you believe them a bunch of rosy-cheeked young babies. On the contrary, they constitute a corps of highly intelligent, rugged, determined, loyal, and fearless young officers. It is my most cherished honor to be their chief, and no human being worthy of the name could betray their trust. Today they have no superiors in the world as military fliers and, given a fair chance, I am certain they will, in a very short time, be carrying the air-mail, in a favorable comparison with their excellent commercial pilot comrades. Our military pilots are not weaklings looking for sympathy! They should be treated like men and not children. If there is anything that the average red-blooded American military pilot resents, it is this recent twaddle about inexperienced, rosy-cheeked boys being sent to their deaths. They are proud that theirs is a man-sized job, and they have no desire to be coddled. Such public statements are a direct challenge to the fighting spirit of the average young Army pilot, and detrimental to the iron flying discipline so essential to a military

> pilot. Therefore, I urge the thinking public to give the Army air-mail pilots its loyal support. Lack of trust eventually will break the morale of the strongest of seasoned veterans and, when morale is gone, all is lost.

I also pointed out that accidents will probably occur in both commercial and military aviation and that both types of flying "constitute a stern and serious business." I added that I wanted the public to know that the air-mail task presented an opportunity for their Air Corps to improve its organization, aircraft, and equipment, and was an excellent way to train pilots in peacetime for any wartime emergency that may arise.

I concluded by putting in a word from the President that he had instructed me to utilize the services of the experienced civilian air-line pilots and mechanics wherever practicable. In my final sentence I thanked the Columbia Broadcasting Company for "the privilege of speaking to the families, relatives, and friends of the men who are now carrying the air-mail, and again to say to them that every safeguard and every care will continually be taken in their behalf."

I have always been amazed at the way the English language can be twisted out of shape so that a listener does not get the message that the sender intended. I thought my broadcast had been given in down-to-earth language, but the word merchants of the press certainly twisted it. Newspaper editorials condemned me for being so callous and inhumane that I deliberately sent gallant young men to their deaths in flaming coffins. Others accused me of having joined the administration bandwagon and of being a political pawn. Still others believed I had joined the Republicans and was disloyal to my Commander in Chief. It is no wonder that Will Rogers remarked on his regular Sunday broadcast that "after listening to all hands I didn't know no more than I did before."

And so the issue of the Air Corps flying the mail became a political football. Representative Hamilton Fish of New York seemed to be the opposition spokesman on the air-mail contract issue, saying "it is simply a striking example of the dictatorial action of an administration drunk and arrogant with power."* I didn't care about the political issues involved and I was frankly surprised at how my remarks had been bent out of shape. I still saw the mail-carrying job as a military assignment, and it now made no difference to me *why* we had gotten it.

On February 26 I reported to General MacArthur that we had experienced several handicaps which prevented us from maintaining schedules: (1) the exceptionally bad weather on the Northern routes; (2) the necessity for creating an emergency organization to do a job we were not prepared to do; (3) lack of adequate funds; and (4) uncertainty as to how long we would have to carry the mail. I made a number of recommendations including an "immediate increase in the number of mechanics at Air Corps depots to maintain aircraft now in use" and the immediate procurement of new planes.

I had *not* been given a time-planning factor and I did not get the funds I had asked for. Instead of the $800,000 I was supposed to get from the Post Office Department, I finally received $300,000 from various War Department reserves which had been earmarked for other projects. The money shortage was serious, and I had many reports from my pilots to show what the impact was on their living conditions. Pilots and mechanics had to live on the local economy at the control points along the routes. They had to rent rooms at hotels (there were no motels in those days) or private houses. They had no per-diem pay and their hours were long. One pilot, Lieut. John R. Sutherland, told of his first run into New Orleans, and his report was typical:

> I was told that there was a place called the Happy Landings Club nearby. I stayed there that night: The proprietor seemed always to be drunk and, as a consequence, was very hospitable. About four-thirty in the morning I was awakened by a terrible racket coming from under the house, which was built on stilts. This racket continued far into the morning, making sleep impossible. I later found out that it was the geese kept by the family. From that time

* *Congressional. Record.* The "arrogance-of-power" phrase was used again in 1966 when Senator Fulbright cautioned against it in connection with the Vietnam war.

on I slept on a GI folding cot in the operations office.

The enlisted men suffered the most from my not being able to get per-diem funds. They were living entirely on credit and handouts. Many were earning only $17 per month after deductions and had no bank accounts to fall back on. The officers were sometimes able to sign for bank loans on the strength of their ranks and pay scales but no enlisted man could gain that kind of confidence from a banker in those days.

The pilots with checking accounts didn't always make out either. I had ordered all personnel not to wear their uniforms on air-mail duty so all they had to warm the cockles of a banker's heart was an Air Corps identification card. The farther away the officer was from his own bank, the louder the "No!" from the banker. As a result, many a pilot took off on an empty stomach and landed at the next stop with his insides growling as loudly as his engine.

The fact that morale held up shows the kind of men we had (and still have) flying our planes. When I enquired about working hours at the various stations, I found that eighteen hours a day, seven days a week, was the normal schedule. In one case at Atlanta a new pilot reporting for duty asked an older officer about the duty hours. The latter, filling out a mail form, looked up at the youngster and said, "Well, Junior, I'm going into my thirtieth hour now without sleep if that will help answer your question."

There were more problems that should have lowered morale. Supplies were not delivered to the control points on time, and there were not enough men, tools, or facilities for maintenance anywhere on the entire system. Mechanics were repairing planes in howling blizzards and drenching rains with makeshift tools borrowed from local gas stations. Those who could do their maintenance indoors had to do it in unheated hangars and often by flashlight.

In spite of the personal hardships, there was a humorous side. Capt. Caleb V. Haynes told of the time his Richmond station was short a plane to take a South-bound run. A plane arrived from the South with North-bound mail which was promptly put aboard another plane and sent on to Washington. The control officer, who had been on duty too long, had a fixation about being an airplane shy for the South-bound run. He ordered the South-bound mail loaded aboard a mail truck and commanded the driver to "drive like hell to the railroad station to get these sacks on the train." Just as the truck arrived at the Richmond Station, the control officer shouted, "Stop! Turn around and drive like hell back to the airport!"

The control officer had suddenly remembered that he could use the plane that had just arrived from the South for the return trip. When he got back to the field, the pilot scheduled for the return Southbound run had assumed he already had the mail aboard since his was the only plane available and was about to take off. The mail-truck driver never did figure out what was going on and concluded that one requirement to be an officer in the Air Corps was to have a short circuit in the brain somewhere.

One of the classic stories to come out of those air-mail days was about the pilot who did take off without the mail and flew his route with an empty baggage compartment. Another was about the young pilot who safely stowed the mail aboard and took off into the night. A few minutes later he radioed back to the control officer: "Where in the hell is this mail supposed to go? My manifest is locked in the mail compartment."

And then there was the pilot who was so eager to make his return run that he took off before the incoming mail had been unloaded and flew it right back to where it had originated. And saddest of all was the tale of the lad whose laundry got mixed up with the mail somehow and was carried from one coast to the other. It was returned a month later by the Post Office Department along with a postage-due notice of $55.

There is an old pilot's saying that "most emergencies come at the wrong time." The air-mail emergency certainly came at the wrong time for the pilots who had to fly their routes in open-cockpit planes. Temperatures were below freezing at night over most of the entire United States except perhaps the Jacksonville-Miami run. Worst of all were the Middle Western routes where the free air temperatures were recorded at ten and twenty degrees below zero.

As soon as unfavorable reports would come to my attention that deserved prompt action, I sent Oscar Westover out to investigate them. Unfortunately, most of the reports of lack of equipment, clothing, and supplies were found to be true. While we were working on an around-the-clock emergency basis, the Army supply system was not, and apparently the Quartermaster Corps couldn't care less that we were on a wartime basis and battling the flier's immortal enemy, the weather.

Westover reported in late February that conditions were as bad as we had been told and getting worse each day. There was still no per diem for the pilots and mechanics. All personnel were being over-worked and every kind of supply item was short in some measure. The only thing that wasn't bad was morale, and Westover frankly admitted that he couldn't understand how any group of men working under such impossible conditions could stay with the job so long and like it. The only other happy note was that March brought better weather on most routes. The worst seemed to be over.

At this time we were supposed to get new Martin B-10 twin-engine bombers to replace some of the P-12s and observation planes on the routes. The B-10s had enclosed cockpits and retractable gear and could cruise at better than 190 miles per hour. They were the first of the monoplane bombers to reach production and they were coming off the line that spring. It was the ideal plane for the air-mail job, so I announced that they would be put on the routes as soon as possible. I then called fourteen Air Corps Reserve pilots to active duty to fly the same routes they had flown for the civilian air lines. They had all volunteered, and we were glad to have their experience. But others had balked because they were afraid they would lose their airline seniority or even their jobs should the airlines get their contracts renewed. I checked with several of the major lines and was assured that no such action would be taken against any of their employees who might be called up for the emergency. This left me with no alternative. I promptly ordered the commissions canceled for those who would not accept orders to report for active duty.

On March 8 Postmaster General Farley passed instructions to me to plan to continue operations until June. At that time it was expected that the civilian air-mail reorganization plan then under study would be put in effect, and we could turn the operation back to the air lines. This announcement gave us something to go on but it didn't help the men in the field any, so the next day I sent a formal report to General MacArthur complaining about the living conditions of my men. “Lack of ready funds to pay enlisted men the per-diem rate of $5 per day,” I wrote, “has created an embarrassing condition which reflects most unfavorably on the Army. Officers and men are quartered at considerable distances from the fields at which they serve. Temporary credit allowed them during the early days of the emergency has been strained to the breaking point.”

I thought this report would get some action from General MacArthur but he never saw it. It was short-stopped in Major General Hugh Drum's office and bounced back to me with the remark that he “desired report without delay as to what steps are being taken to relieve the situation.”

If Drum had read the newspapers he would have known that the bill providing per diem for our personnel on air-mail duty was still hung up in the Congress where the politicians were arguing over the whole air-mail issue. With this kind of support from Drum, the Air Corps didn't need any enemies. There was nothing I could do about his supreme ignorance except grit my teeth and make a silent vow that I would by-pass Drum whenever I could to get to MacArthur.

On the night of March 9 Lieut. Otto Wienecke, flying alone, took off from Lylertown, Pennsylvania, in an O-39. Two hours and twenty minutes later he lay in its wreckage near Burton, Ohio. He had flown into a heavy snow squall and apparently spun straight into the ground. The exact cause was unknown but it could have been inoperative gyros, ice on his wings, engine failure owing to carburetor icing, or a case of vertigo. All we knew or would ever know was that he was dead.

That same night, Lieut. W. W. Reid was taking off from the Daytona Beach, Florida, airport in a twin-engine B-6 when he lost an engine and smashed into the woods adjoining the field. Reid and a passenger, Pvt. A. M. Marshall, leaped out unhurt but the mechanic, Pvt. E. B. Sell, died on impact.

As if two fatalities on one night were not enough, within minutes of learning of the first two I received a radiogram from the Western Zone reporting the death of Lieuts. F. L. Howard and A. R. Kerwin on a night takeoff from Cheyenne, Wyoming. Again we never found out what really happened, but their deaths could have been the result of a too-heavy load trying to take off from a high-altitude airport. The O-38E they were flying was underpowered and, as Arnold described it in his report, "very unpleasant… to fly at high altitude."

On the morning of March 10 I received a call from General MacArthur to accompany him to the White House. When I got to his office, he was pacing the floor.

"Benny, you're in trouble," he said. "I don't know how you're going to explain three fatal accidents in different parts of the country in one night, but I think you had better try to find the words somehow. The President is extremely upset by the public reaction to these three deaths."

I nodded and swallowed hard. How do you "explain" aircraft accidents to a man who has never flown and how do you communicate the dangers inherent in flying through the worst winter weather the country had seen in over half a century? Is it possible really to "explain" the difficulties of maintaining a plane in the open without tools, the problems of a pilot flying instruments alone at night without adequate clothing, or the uncertainties of landing a hot pursuit plane in a strong cross wind at a poorly lighted airfield? Are there the right kind of words in our language to describe the personal difficulties of men who exhaust themselves without adequate compensation or tell about the physiological problems a man has when flying on an empty stomach at high altitude without proper rest?

MacArthur and I drove to the White House in silence. I had the feeling that a condemned man must have when he is being taken for that last walk. When we arrived at the White House, Stephen Early met us at the door and escorted us upstairs to the second floor. MacArthur whispered to me that the President was probably still in bed.

Early ushered us through a set of double doors into a huge bedroom. Except for its size it was an ordinary bedroom with the usual dressers, chairs, and personal effects of the occupant. But this was the bedroom of the President of the United States and there, propped up in the famous Lincoln bed, was my Commander in Chief looking like a high potentate ready to view the prisoners who had committed crimes against his kingdom. The dark wood of the bed contrasted sharply with the white covers and the canopy that hung from above. There were newspapers piled on the bed and sheaves of State Department messages and reports piled in small stacks on both sides of him.

MacArthur was a stride ahead of me as we approached the bed. “Good morning, Mr. President,” he said. “This is General Foulois, chief of the Air Corps.”

I felt a certain awe in being in the most famous house in America and being introduced to the most famous man in America. But the man I saw wasn't the man whose picture was everywhere. The familiar smile was gone. In its place was a deep scowl which was fixed directly on me. Without a word of greeting the voice everyone had come to know so well boomed, “General, when are these air-mail killings going to stop?”

If that question was calculated to throw me off balance, it certainly did. But my answer came back without me being fully aware that I was saying it.

“Only when airplanes stop flying, Mr. President,” I replied in a calm voice which surprised me.

There was a deep silence as the President stared fiercely at me through his pince-nez glasses.

“I've heard nasty rumors that there has been a premeditated

propaganda campaign launched by the airlines to embarrass the Air Corps," he said. "Even sabotage has been mentioned. What do you know about that?"

"I am well aware that the press has been instructed to belittle and embarrass the Air Corps, Mr. President," I answered, "but I have not heard about any sabotage, and I don't think anyone connected with flying would deliberately send another man to his death by tinkering with his machine."

"Well, believe it or not, General, I want you to investigate those rumors and put a stop to these killings."

For the next ten minutes MacArthur and I received a tongue-lashing which I put down in my book as the worst I ever received in all my military service. There was no doubt that what bothered Roosevelt the most was the severe criticism his administration was getting over the contract cancellation. He did not seem genuinely concerned or even interested in the difficulties the Air Corps was having. It was the bitter editorials, the anti-administration slant of the news stories about the three accidents, and the quotes from Capitol Hill opponents that had made him angry. The fact that three brave men had just died trying to do his bidding in the wrong equipment and under the worst possible weather conditions did not seem to concern him. That was my fault, he said. What was worse, and also my fault, was that my personal shortcomings were misdirected at him and Farley, and this he highly resented.

He waved a piece of paper through the air as he spoke and then told me that he had just dictated a letter to Secretary of War Dern directing that "the continuation of deaths in the Army Air Corps must stop." He wanted additional safety precautions to be taken even at the expense of reduced service and ordered the Secretary to take the "utmost care" to prevent "constant recurrence of fatal accidents."

MacArthur said nothing all this time, and for the first time in my life I sensed the wide gulf that exists between the man in uniform and the transient politician who is suddenly placed in a position of the highest authority. The military man trains all his adult life to attain the next higher rank and position. The politician may be a small-town banker or lawyer, yet by the mere accident of making friends with the "right" people or garnering enough votes can be catapulted into a position in our government where, with a snap of his fingers, he can send a military man to his death or end his career. Here was one of those politicians who did not understand air power, airplanes, or any of the problems of flying and apparently did not want to learn. From the tone of his voice and his derision of the Air Corps I was, to him, just another of those crazy, undisciplined fly-boys who had progressed in rank far beyond my capabilities. He mentioned "those killings" several more times, and I felt a strong resentment toward the man and his harsh words. When I tried to explain, my words never got beyond my teeth. The leading character in this drama was delivering his monologue and the other actors were only supposed to stand by and listen.

MacArthur and I were finally dismissed with a curt wave of the hand, and I left the White House that day completely dejected. The only guidance I had received was to "stop those killings." The inference was that I had somehow personally ordered the three unfortunate lads to their deaths, and that my negligence and mismanagement had caused these tragedies to be laid at the White House door. There was no doubt in my mind from that day forward that one Benjamin D. Foulois was to be the "heavy" in the air-mail drama.

When we arrived back at the War Department, I started to explain to MacArthur some of the things I had put in my report but which he had not seen. He didn't want to hear about any of my problems, either. "Benny," he told me, "I told you that this was your show all the way. That means you have to take the brickbats as well as the bouquets. You've just gotten your brickbats. The only thing you can do is get back to work and do the best you can. I'll back you when I can but I can't win your battles at the White House for you, as you can see now. We've got budget hearings coming up and I'm sure the Republicans will give us the chance to fire a few rounds for effect. Make sure you have your ammunition ready when you get the chance to fire."

The letter to Dern was a calculated move to smooth the feathers of confused Democrats and answer the Republicans. It was given wide publicity and was calculated to prove to the public that the President had taken the necessary action to place the blame precisely where it belonged—not in the White House but in the office of the chief of the Air Corps. I was the sacrificial goat. As a military man, I had to take this dubious honor without complaint.

While I felt duty-bound to keep my thoughts to myself, my men in the field did not. They sought out the newsmen and spilled their feeling about the President's lack of knowledge about the safety measures in effect and pointed out that only four of the deaths had occurred on regular mail runs. They were frank about their personal difficulties with finances, the long hours, and shortages of supplies. Their frankness got many inches of newspaper space and the controversy built toward a climax.

The next morning, on March 10, I called my staff together and told them I had decided to ask for a reduced route structure and was ordering all operations to cease for ten days while we laid out the new routes. I then met with postal officials and we agreed on nine routes with a daily mileage total of about 25,000 miles.

During the stand-down period I sent orders to the three zone commanders to install and test two-way radio sets and a complete set of instruments in all planes to be used for night runs and the same in as many day-flying planes as possible. All pilots with low flying time were to be returned to their units and replaced by the most experienced pilots available. In the meantime, I hopped in my plane and visited as many of the principal mail stations as I could to talk to the pilots, check their equipment, and see what I could do to ease their problems. Just as Westover had found, the morale of the officers and men was the highest I had ever seen of any unit in the service. They still had no per diem, some of them were working under the most primitive arrangements imaginable, and many were gaunt from prolonged physical exhaustion. Only the knowledge that they were doing a job that someone in authority had said needed doing kept them going. The tougher they had it, the more determined they were to show they could master the situation.

We had planned to resume operations on March 19. Two days before, however, Lieut. H. C. Richardson, one of the airline pilots recalled to active duty, crashed and was killed in an O-38B while on a training flight near Cheyenne. Ironically, the news of this crash reached me just as I concluded a radio broadcast in which I told the public about the new safety measures we had just put into effect.

The President's letter to Secretary Dern plus rumblings of a new congressional investigation of the Air Corps and the death of Richardson prompted Dern to order his own investigation. He invited Charles Lindbergh, Orville Wright, Clarence Chamberlain, and the members of the Drum Board, including me, to serve on the study group. Lindbergh refused to serve, saying that the use of the Air Corps to carry the mail "was unwarranted and contrary to American principles." Orville Wright could not serve because of illness, so Dem sent invitations to Newton D. Baker to serve as chairman, Jimmy Doolittle, Edgar S. Gorrell, and Dr. G. W. Lewis.* This committee was asked to study all phases of military aviation and make recommendations to Dern as to how the Air Corps could be improved.

Operations resumed on March 19. The weather improved and the new B-10 bombers were arriving from the factory almost daily. The reduced schedules and route mileage enabled us to get going on a more relaxed basis but there were still personal problems for the men on the routes. They still had no per diem, and the total amount owed to local townspeople by Air Corps personnel across the country was estimated to be about $250,000. Finally, on March 27, the emergency air-mail bill passed and was signed by the President authorizing the Post Office Department to transfer funds to the War Department. Needless to say, there was great rejoicing

* Jimmy Doolittle, a Reserve officer, was working for Shell Oil Co. after having resigned his regular Air Corps commission in 1930. Gorrell, a former pilot during World War I, had become president of Stutz Motor Co. Lewis was director of research at NACA.

in many towns across the land as officers and men cashed their government checks and paid their debts.

There was another bright note, however, that received no notice. The pressure of criticism about the air-mail had caused the President to release Air Corps research and development funds that he had impounded, and he even authorized $7 million in PWA funds to be used for new aircraft and other badly needed equipment. This apparent change of heart, however, had been inspired by political reasons and not by any great wisdom or hopes for the future of military aviation.

Things looked a lot brighter on the horizon now. The B-10s were placed on the main runs and all mail pilots were ordered through an instrument course at Wright Field conducted by Capt. Albert F. Hegenberger. The fine art of instrument flying became uppermost in our thoughts, and I encouraged the development of new techniques and instruments at every opportunity. Our hope was that a rugged, dependable system could be devised that would enable a pilot to bring his plane down to the point of flare-out for landing. It would be several years before two basic systems (ILS and GCA) were available, but the research for them actually began under our encouragement in 1934.

The final fatal accident on the air-mail operation occurred on the night of March 30 when Lieut. Thurman Wood, flying an A-12 was approaching Davenport, Iowa. He flew into a severe thunderstorm and in trying to reverse his course lost control and spun in. It was the same combination of night and bad weather conditions that had been the major contributing factors in other accidents. There was only one answer—more training in both instrument and night flying and the proper equipment to conduct the training.

During April, it appeared certain that we would not have to carry the mail beyond June 1. But with the funds available I pushed the zone commanders to get as much pilot training as possible accomplished in addition to flying their routes. Once we quit flying, I was afraid we might have to revert to our frugal existence again and operate with ever-diminishing War Department funds. At the end of April we were instructed to phase down all operations by mid-May. When word got around the Circuit, several of Hap Arnold's pilots wanted to see what they could do with the B-10.

All of us had been rankled by the publicity Eddie Rickenbacker had gotten for himself when he had flown the last commercial plane mail East in a little more than thirteen hours. On May 7 the last Air Corps coast-to-coast mail run began in San Francisco. Fourteen hours and eight minutes later the B-10 delivered the mail in Newark. Although the time was not less than Rickenbacker's, we all felt a moral victory had been won for the Air Corps. Our pilots had made three more stops than Rickenbacker had and their total distance was 279 miles more.

All but one of the routes phased down by May 17. The only one that did not revert to civilian air lines at that time was the run between Chicago and Fargo, North Dakota. However, on June 1 the last mail pouch was delivered by an Air Corps pilot and we closed the books on a tragic but vitally important chapter in the history of American military aviation.

After it was all over, Postmaster General Farley, in a letter to me, predicted that "the country and the Congress will, without doubt, give a more adequate support to the Army Air Corps, will see to it that it has the most modem equipment obtainable, and that sufficient funds are provided for the fliers to have the additional hours of flying which have so long been needed." He thanked us for our commendable service and noted, "Had it not been for the Army Air Corps the country would have been without any air-mail service for a period of more than three months. It is a notable fact that not a single pound of mail was lost during the time the Army had flown the air-mail."

The Air Corps phased out of the air-mail business without many other words of praise, but I have always believed our pilots did a magnificent job. They logged more than 13,000 hours of flying time, flew more than 1½ million miles, and carried more than 777,000 pounds of mail. They set a number of speed records over their routes. But it wasn't their successes that contributed most to progress. It was the failures, because the failures focused public attention on the inadequacies in our training and equipment—inadequacies caused by lack of Congressional interest and funds. The sudden need to provide a national service and the subsequent tragedies proved how weak our air power was as nothing else could have short of war.

It was from this tragic experience that the first giant step was made toward the creation of an independent Air Force. While the air-mail operation was being phased out, the Baker Committee, of which I was a member, met for a month beginning on April 17. We heard 105 witnesses, including two former assistant secretaries of the Army and Navy for air, generals, admirals, noted civilian pilots, airline pilots, and specialists in engines, instruments, blind flying, and weather. The questioning covered the entire spectrum of military aviation, and the sessions, under Newton Baker, were conducted with great decorum. The only loud voice of dissent on the committee was Hugh Drum whose antagonism toward flying and fliers was so obvious that I almost felt sorry for him. He knew so little about airplanes that I wondered how a man could sit at the elbow of the Army's Chief of Staff as he did for so long and be so ignorant.

After twenty-five days of testimony which was recorded on 4,283 pages of the record we concluded our hearings and adjourned to write our report. The Baker Committee was the fifteenth such group to be convened in sixteen years to study military aviation and it was our hope that our committee could finally lay the basis for its proper development for the next decade.

I fought many verbal battles during our closed-door deliberations. My chief opponents were the four ground generals aligned solidly against me. Jimmy Doolittle, a master at listening until the right psychological moment for a verbal touché, would consistently clinch an argument with a summary so full of logic that he would have my opposition agreeing with me and not even realizing it.

It took us until the middle of July to come up with our conclusions and recommendations. The report, unfortunately, reflected the thinking of the highly opinionated Drum whose mind was continually closed to the potential of the airplane. It was majority opinion that no basic change in organization of the Army and Navy was necessary, which meant that there would be no separate Air Force and that the civil and military aviation functions should not be separated.

What disgusted me most was the fact that the report denied most of the claims for the potential of air power by stating that "independent air missions have little effect upon the issue of battle and little upon the outcome of the war." Although I objected violently to this statement, I decided to give in because the committee did recommend creation of a GHQ Air Force made up of air combat units and capable of operating either independently or in cooperation with the ground air forces. This was a compromise and not exactly what we wanted, but it was a step forward and that was better than no step at all. I signed the report and was anxious to get on with planning for the future using the good points of the committee's recommendations as my rationale for proceeding.

Jimmy Doolittle decided to present a minority statement which he had every right to do. After I read it, I wished that I had joined him in making it. He said:

> I believe in aviation—both civil and military. I believe that the future security of our nation is dependent upon an adequate air force. This is true at the present time and will become increasingly important as the science of aviation advances and the airplane lends itself more and more to the art of warfare. I am convinced that the required air force can be more rapidly organized, equipped, and trained if it is completely separated from the Army and developed as an entirely separate arm. If complete separation is not the desire of the committee, I recommend an air force as a part of the Army but with a separate budget, a separate promotion list, and removed from the control of the General Staff. These are my sincere convictions. Failing either, I feel that the Air Corps

should be developed and expanded under the direction of the General Staff as recommended above.

It was to be thirteen years before our dream and Doolittle's recommendation would be carried out. It would take a war to prove that air power could, indeed, produce decisive results—a war that would end with a single plane dropping a single bomb and thus forever altering the course of history.

experience in aircraft manufacture, their financial status, or plant capability. The plain truth is that if the Air Corps had attempted to buy its aircraft on this basis, there would have been no Air Corps. Low bid does not mean competence. We needed planes, and we knew who could build them. We published our specifications and challenged the aircraft industry to build planes to meet them. If they did, we negotiated a fair price based on what their planes cost plus a fair profit. If two or three manufacturers built planes that could meet the specifications, we usually purchased a few of each for service test to see which was best. The best would then get a production contract.

The fight over the aircraft procurement dollar became serious as the Roosevelt administration cut the defense budget to make way for its pump-priming programs. When the 1933 budget was made up, the Army had been ordered to take a $15 million cut. Supposedly, this was to be distributed evenly throughout the Army. But the General Staff saw to it that the Air Corps was given a direct cut of $5 million and an indirect cut—through quartermaster's supplies, housing construction, etc.—of $6.5 million. The next year, the Congress appropriated more than $8 million for new planes but the War Department budget officer, under the control of the General Staff, permitted the expenditure of only $1.3 million. The balance was impounded in the Treasury. As a result, we had only about 800 serviceable planes by 1934 instead of the 1,800 we were supposed to have.

And so it was that while the Air Corps grew more obsolete daily, the funds that might have helped to reverse the situation were withheld by the General Staff. However, in late 1933 the Public Works Administration was authorized to transfer $7.5 million to buy planes, parts, engines, radios, and machine guns. I had appeared before the House Appropriations Committee five days before we started flying the mails to tell how I intended to spend the money. We were going to buy 105 planes from the Northrop, Boeing, and Martin companies, I told the committee, and we knew exactly which planes we wanted and how much we could buy them for. This was considered "a negotiated purchase." At this point Assistant Secretary of War Harry H. Woodring directed that all such purchases had to be made by competitive bidding.

On May 7, 1934, the Rogers subcommittee published a report of its investigation which said that my intent to buy the 105 planes was "in clear violation of the law." Woodring was immediately dubbed the hero in this new drama for insisting on competitive bidding and stopping the transactions. The only way Woodring could achieve his purpose, however, was to force us to lower our specifications for the planes we wanted by reducing speed, range, and altitude requirements. This I refused to do. It would have meant buying new but obsolete planes.

During four months of hearings the subcommittee called Oscar Westover, Conger Pratt, and others assigned to my office as witnesses. I appeared three times before the Rogers Subcommittee and each time was subjected to a severe cross-examination where my guilt was not the question in the minds of the committee—only its extent. In June another report was released by the committee which accused me and my subordinates of deliberate and willful violation of the law. I was personally cited for mismanagement and inefficiency in the handling of the air-mail. The report concluded with the statement that "We are most firmly convinced from the evidence and records submitted that before any substantial progress in the upbuilding of the morale and the matériel of the Army Air Corps can be attained, Maj. Gen. Benjamin D. Foulois must be relieved from his position as chief of the Air Corps. We unanimously recommend that the Secretary of War take such action without delay."

If I had been charged with fraud against the government in a general court-martial, I would have been able to defend myself against the charges. But the "trial" which had been conducted by the Rogers Subcommittee in which I was the principal accused had been held in closed sessions at which I was not present. I was not permitted to see or know about the evidence against me. The "jury" had deliberated, returned with a verdict of guilty, and sentenced me to be ousted from the service.

What were the facts? What would I have said if I had been allowed to speak in my own behalf?

I could have made several points clear beyond any doubt. I would have shown that the procurement of planes by direct negotiation with the manufacturers was followed by General Patrick from 1926, was continued by General Fechet until 1931, and continued by me until Secretary Woodring called a halt in 1934. I would have shown that every negotiated contract had been submitted to the Judge Advocate General of the Army for approval and that each had been declared "legally sufficient." I would also have shown that the Secretary of War had written "approved" on the face of every contract and signed his name. And finally, I would have reminded the "prosecution" that under the law the chief of the Air Corps could only *recommend* approval and had no final approval authority. In my defense, I would have asked the "court" who is guilty of violating the law—the chief of the Air Corps who *recommends* approval, the Judge Advocate General who passes upon the legal sufficiency of a contract or the Secretary of War who finally makes the contract a legal document by signing his name to it?

A man accused of crimes he did not commit must endure a living hell inside himself. What can he do? I saw only one thing to do: fight back.

While the Rogers Subcommittee was investigating my alleged mismanagement and inefficiency in handling the air-mail, it was also carrying on a conflict of interest investigation in connection with lobbying practices, the purchase of motor vehicles, and the disposal of surplus War Department property, which involved certain members of the Quartermaster General's Office and a member of the Judge Advocate General's Office.

When this conflict of interest investigation was publicly advertised by the Rogers Subcommittee, General MacArthur called me to his office and said, "Benny, have they got anything on you?" To which I replied, "Not a thing, Doug."

I then briefed MacArthur on my then thirty-three-year-old

policy of handling so-called conflict-of-interest cases. I also briefed him on numerous gifts which I had received from aircraft manufacturers while chief of the Material Division (1929-1930)—gifts which I immediately returned, accompanied by formal letters, copies of which were in my files, to the effect that as long as I held office as a public official I looked upon my job as a public trust which barred me from accepting gifts of any kind.

I further briefed him regarding certain information, which had been transmitted to me by a friendly member of the House Military Affairs Committee, to the effect that the Rogers Subcommittee had already tried to pin a conflict-of-interest case on me which had backfired unexpectedly. This case arose during the questioning of one of the star witnesses for the prosecution—an aircraft manufacturer who felt that he had been unjustly treated by the Air Corps in awards for aircraft contracts. In his testimony before the Rogers subcommittee he was asked by one of the members of the committee if he knew of any case in which General Foulois had taken gifts or other favors from aircraft manufacturers. His answer was short and to the point:

"Hell, you can't give that man a five-cent cigar!"

Having fully satisfied MacArthur as to whether they had anything on me, I then suggested that he endeavor to keep the President and the Secretary of War off my back and I would take on the Rogers Subcommittee—an agreement which continued from that time on.

When the Rogers report reached the public through the newspapers, the reaction was interesting. The headlines on the front pages that day said:

LAW VIOLATION LAID TO
FOULOIS, AIR CORPS CHIEF

FOULOIS HELD LAW
VIOLATOR

FOULOIS GUILTY ON PLANE
BIDS, COMMITTEE SAYS

For the most part, the reporters accurately repeated the accusations in the Rogers report without slanting their stories. However, the inference in all of them was that I was guilty simply because the committee had spent so much time investigating the circumstances and the committee was made up of honest, truth-seeking representatives of the people. I was asked for interviews by the Washington press corps and I refused. About a week later several correspondents called on me and one of them said, "General Benny, remember how a few weeks ago we told you Vie had orders to 'get' you when you started flying the mail?"

I nodded.

"Well, it might interest you to know that we have just gotten orders to go all out to support you!"

And support me they did, much to the chagrin of the politicians on both sides of the aisle on Capitol Hill. Laudatory editorials appeared in newspapers and magazines all over the country supporting me. Researchers dug up the embarrassing information that the Military Affairs Committee had known of and voiced no objection to the theory of negotiated contracts since General Patrick had made the first one in 1926. Then diligent reporters found out that the subcommittee had held many of its meetings behind closed doors in so-called "executive sessions." When they asked for transcripts of the testimony given by secret witnesses, they were refused. There were persistent rumors that parts of the testimony had been burned, lost in the mail, or otherwise destroyed. The tone of the articles and editorials changed dramatically in favor of the Air Corps and its chief. One writer who had accused me of "legalized murder" a few weeks before wrote praiseworthy stories saying that I was "a public servant deserving of the gratitude of the nation."

The eight members of the Rogers Subcommittee must have squirmed uncomfortably in their plush chairs as they were accused of attempting to run an innocent man out of the service by their star-chamber proceedings.* As the circumstances surrounding my "trial" were researched and published, I began to receive "fan mail" from people all over the country offering me their

assistance. American Legion posts and civic groups passed resolutions and wrote scathing letters to their congressmen. I received requests from magazine and newspaper editors offering to pay me huge sums if I would “tell all”.

As the days went by in the summer of 1934, the press persisted in its efforts to vindicate me. In retaliation, the subcommittee insinuated to Secretary of War Dern that if I were not relieved, he could expect no additional funds for the expansion of the Air Corps.

Secretary Dern must have had some difficult days after he officially received the request from the Rogers Subcommittee to have me relieved as chief of the Air Corps. And I don't suppose his nights were made more cheerful by the editorials and news stories in the evening papers that made me out as the underdog in the game that had been played with loaded dice. Dern never did discuss the matter with me at any time, although my office was not far from his. Instead, he transmitted the report and its recommendation to me for such reply as I cared to make.

I worked on my rebuttal for six weeks and in mid-August, 1934, sent a long, preliminary statement back to Secretary Dern explaining the facts, pointing out the errors in the “prosecution's” case, and requesting that I be permitted to examine the evidence against me. I pointed out that I had formally requested permission to examine the testimony of witnesses against me three times but that my requests were not even acknowledged.

On August 21 Secretary Dem sent a letter to Congressman Rogers which must have been an agonizing one to put together. Dern was a Cabinet member and highly sensitive to the political winds that blew down from Capitol Hill. But to his credit he composed this classic which I repeat here in part:

* The eight members were William N. Rogers, chairman of New Hampshire; Lister Hill of Alabama; Numa F. Montet of Louisiana; Dow W. Harter of Ohio; W. Frank James of Michigan; Edward W. Goss of Connecticut; Charles A. Plumley of Vermont; and Paul J. Kvale of Minnesota.

A most difficult problem now confronts me, the solution of which involves some of the fundamental precepts of our system of jurisprudence embodying many of the most sacred rights of American citizenship.

As a matter of course, any recommendation of a committee of Congress would receive my most respectful attention. Our system of government requires it. Proper coordination demands it. Had the report of the committee been confined to an indictment or charges against Major General Foulois, the situation would have presented no difficulty. In such event I could have followed the regular procedure prescribed for the determination of such matters by referring the case to a military tribunal for adjudication. Instead, however, the report is not limited to an indictment, but in effect finds the accused guilty, fixes the sentence and calls on the Secretary of War to execute it. Under such circumstances should I now refer the matter to a military tribunal for the determination of the precise matters which your committee already has adjudged, such a step well might suggest a reflection on the committee. Such a suggestion I desire most scrupulously to avoid. My respect for the members of the committee is only exceeded by my desire to work in complete harmony with them.

Accordingly, had the report and recommendation of the committee been based on a full and complete hearing wherein the constitutional rights of the accused were assured, even though the procedure were somewhat unusual, I would have felt constrained to comply with the mandate of the committee.

But such was not the case. The hearings were conducted in the usual manner of congressional hearings held in executive or closed session. Major General Foulois appeared before the committee, but it was in the capacity of a witness and not of an accused. I understand that he was later given an opportunity to appear again before the committee before the report was submitted, but for some reason this did not eventuate. It does not appear that he was informed of the accusations contained in the report, nor was he confronted by the witnesses against him and given an opportunity to hear their testimony and cross-examine them, and then to offer evidence in his own defense. Nor was he permitted to be represented by counsel of his own choosing. All of these rights are sacred to every American citizen and are guaranteed by the Constitution. Major General Foulois states that he repeatedly requested that he be furnished with a copy of the full record of the evidence which formed the basis of the report and recommendation and that this request has not as yet been granted.

This situation places me in a most difficult and embarrassing position, because in the first place the accused has not had full opportunity to examine or to refute the evidence against him; second, I, myself, do not know what this evidence is, except to the limited extent it is set forth in the report; and finally, I have not been called upon to evaluate the evidence or make any decision in the case, but to the contrary am, in effect, precluded from following such a course, due to the findings and recommendations of the committee.

Being convinced that the committee is as solicitous as I am that complete justice be done in this case, I feel that the only step now open to me is to transmit to the committee the partial statement of Major General Foulois to the end that the committee may consider whether he should not now be permitted to have full access to the evidence against him and then appear before the committee to present his defense.

With further assurance of my deep respect and of my desire to cooperate in the solution of this case to the fullest extent consistent with the dictates of fairness and justice, I await the committee's further action.

Sincerely yours,

(Signed) GEO. H. DERN,
Secretary of War.

There was a long silence on Capitol Hill after Dern's letter was received but I began to hear rumblings around the War Department that the Rogers Subcommittee, embarrassed by the unfavorable newspaper publicity over my recommended ouster, had shifted their investigations to a study of lobbying practices, the purchase of motor vehicles, and the disposal of surplus War Department property.

There is no doubt in my mind that the politicians wished they had never made an issue of the air-mail or the method of obtaining planes through the method of negotiated contracts. Between June and December, 1934, thousands of words were printed in the national press supporting me and condemning the Rogers Subcommittee. Secretary Dern, by his masterful reply, became an embarrassed hero. I was vindicated in a backhanded way and was not relieved of my job. I was the acknowledged defender of the negotiated contract which was at that time the only way to get airplanes that would equal those of the other world powers. But while I had won the battle, I lost the war, for when I was forced to advertise for bids for Martin B-I0 and B-12 bombers, the Air Corps lost eight months on the delivery of planes and the taxpayers had to pay several hundred thousand dollars more because of increased prices when we finally did get them. The B-10 was a good airplane—so good that it won the Collier Trophy for Glenn L. Martin. To further prove its value, I decided to show that it would do everything we said it should do. I put in a call to Hap Arnold at March Field, California.

"Hap, I've got another job for you," I told him. "Can you get a squadron together for a trip to Alaska?"

There was a short silence on the other end of the line, then the characteristic Arnold reply, "General, I hear the fishing is great in Alaska now. When do I leave?"

I didn't know it at the time but Hap and his wife had started on a fishing trip of their own to Jackson Hole, Wyoming. Hap had earned the leave after the air-mail operation had phased down, but he left immediately for Dayton where he met Oscar Westover and Maj. Hugh Knerr, then in charge of bomber development. Knerr was made Arnold's executive officer and the two of them began planning their flight. A month later they led ten B-10s to Bolling, got a sendoff with a lot of fanfare, and arrived in Fairbanks, Alaska, without incident on July 24. There they established a base and photographed some 20,000 square miles of Alaskan territory.

The return flight on August 17, made nonstop from Juneau to Seattle, was the first mass flight made between Alaska and

the continental United States without flying over Canadian territory. They continued on to Washington where Hap made his report to me and then took his unit back to March Field. They had flown 18,000 miles over the roughest flying country in North America and had averaged 820 miles per day. Here was proof that our judgment about the B-10 had been right—regardless of the method of procurement used.

While Hap and his twenty-man crew were making air history, another board was formed in Washington. It was an Aviation commission appointed by the President to study the air-mail system to determine what the government should do to encourage air transportation and to chart a new course for civilian and military aviation. Clark Howell, Atlanta publisher, was appointed to head the board and had unlimited authority to investigate any phase of aviation he desired. When he met the press after being sworn in, he declared, "We come to this task with open minds. There will be no pussyfooting, no star-chamber proceedings. All interested parties will be given fair and full hearings."

My reaction at first was negative. With the Baker Board and the Rogers Subcommittee investigation going on, all I could see was another year lost in building our fighting force while the politicians "studied" the aviation picture. But the Howell Board turned out to be more objective than any of the previous fifteen investigating boards, and from it came the Civil Aeronautics Board and the Civil Aeronautics Authority (now the Federal Aviation Agency). Order and organization came to civilian aviation. The funds for airports, stricter safety requirements, and improved navigation facilities came into being and all of this began in 1934—a year too many air historians have painted black.

What the Howell Board was to commercial aviation, the Baker Board was to the Air Corps. In December 1934, the independent GHQ Air Force was officially authorized by Secretary Dern and Oscar Westover was appointed its chief. He was to report only to the Army Chief of Staff, General

MacArthur—not to me. I would be responsible for research, development, and procurement. Of course, I would still have control over training and have many flying units under my jurisdiction. But the first major step toward a separate air force was achieved. Its immediate purpose was to provide a mobile combat air force coordinate with four ground armies and capable of strategic offensive bombing as well as tactical operations in support of ground troops. We didn't discuss its long-range purpose—the real reason we were happy with the Baker Board recommendation. That reason was simple: we could now concentrate on building an air force to support this concept.

Just when I thought that all of the battles were over, the Rogers Subcommittee arose again and issued a statement to the press saying they were going to renew their investigation into the "vicious and improper practices" of the Air Corps. Among other accusations, Chairman Rogers charged that the formation of the GHQ Air Force was illegal and that the Air Corps had no right to plan an air force that could fly beyond the coast to repel enemy planes or vessels.

Rogers and his colleagues blamed me for the decision to create the GHQ Air Force and again demanded my ouster. This time they had additional ammunition. Secretary of War Dern, under continual pressure from the Democrats, had directed the Army Inspector General to investigate me. The investigation was completed by June, 1935, and Dern reported to the Rogers Subcommittee that "the evidence adduced did not establish that General Foulois violated existing laws in the purchase of airplanes and aircraft materials" and "that there was no cause for censure in General Foulois' optional statements with regard to the capacity of his corps to fly the air-mail…." But the Inspector General added one more statement which said that I "did depart from the ethics and standards of the service by making exaggerated, unfair, and misleading statements to a congressional committee."

When the Army Inspector General's report reached MacArthur's office, he called me in and said, "Benny, I'm going to

stop all of these investigations if possible. I'm going to reprimand you for using unorthodox language against the War Department General Staff."

I looked at him with a grin and said, "You have a good case, Doug. I've been using unorthodox language against the War Department General Staff since 1908. I have been subjected to sixteen major investigations since that year, and I received a moral wound stripe in the form of reprimands for every one of them. One more reprimand won't be too much of a burden if it will stop this present investigation and let us get on with the development of the Army Air Corps."

MacArthur's "reprimand," approved by Secretary Dern, ended the investigation—that is, insofar as the War Department was concerned. However, Rogers was still on the warpath. In a bitter speech on the House floor, Rogers branded me a liar and a perjurer, saying that Secretary Dern had merely given me a slap on the wrist. At this point Republican members leaped to my defense, and a verbal exchange took place in the House chamber that blistered the furniture. The more the Republicans tried to shout him down, the more vitriolic Rogers became until he finally shouted that I had to be relieved immediately or else. He also charged that Dern had given the press a misleading statement saying that I had been cleared "except for a little unethical testimony."

The "wrist slap" that Dem and MacArthur had given didn't bother me, but what did was that the dispute about my staying on the job was holding up the budget for the Air Corps for the next fiscal year. Since we had been stopped from going ahead with plane purchases, we were going to have even less planes then we had the year before. We were more than 1,000 planes short of our announced inventory goal at this time. The Rogers Subcommittee, with all its shouting, was damaging the Air Corps irreparably and that made me sick with frustration. All we had worked and prayed for was slowly going down the drain. What made our plans for expansion all the more desperate was the announcement by Herman Goering, the same Goering I had befriended in 1920, that he was re-creating the German Air Force. On March 16, 1935, he boasted to the world that the Luftwaffe had been reborn and would be the strongest air force in the world.

Goering's announcement was certainly no surprise to me. I had learned through intelligence channels that, in spite of Allied inspection teams, the Germans had been making pursuit planes in widely dispersed assembly plants and that their high-altitude transport planes were really bombers that were much better than our B-10s. While the leading airmen of the United States scoffed at the now rotund Luftwaffe chief and his incredible egotism, I could not. He had told me in 1920 that the Luftwaffe would rebound and he told me he would personally lead it to victory. He had bragged that no Allied control commission could prevent the inevitable rebirth of German air might and that he would build his planes right under their noses!

I recalled those many hours in Berlin when the forceful Goering and the quiet Udet would drink my Scotch and tell me step by step how German air power would one day dominate Europe. And then he would laugh heartily and say, "After ve pay back the French and the British, then ve pay you Americans back, eh, Udet?"

Udet would never answer but he didn't have to. Goering would make the plans and Udet would be his quiet chief air engineer. It gave me cold chills to read the intelligence reports and watch the blueprints of the next war being drawn exactly as I had been told they would be by the very architects of that war.

The feeling that war was inevitable overwhelmed me, but the more I read about the demands of the Rogers Subcommittee for me to retire, the more I began to feel that the only way the Air Corps was ever going to get the planes to meet the growing threat in Europe in the person of Adolf Hitler was for me to get out of the way. After much self-debate, I requested and was granted a three-month leave of absence in the fall of 1935 to expire a few days before I would complete my four-year tour as chief of the Air Corps. I would then retire and pass the fight on to my successor.

General MacArthur had a solemn look on his face as I passed the papers across his desk after I had made my decision. "Benny, I hate to see you go," he said. "As far as 1'm concerned, you did a magnificent job, but we both know you are a casualty of political warfare—the kind of war that the military man abhors and cannot win. I have had only one fear in my years of service and that is that I will fall on the battlefield of politics instead of beside my men on the battlefield of honor.

"I will approve your request for leave gladly but I put my signature on your request for retirement with a great deal of personal sadness."

The man who would fight two more wars and go down in history as one of our country's greatest military commanders rose slowly from his desk and extended his hand to me. He gripped mine tightly and closed his other hand on top. As we slowly walked to the door, he didn't let go. For a moment I sensed the fear he had expressed as if he had a premonition of some kind that he, too, would become a political casualty. If he truly had that premonition, it took almost two decades to come about. When it did, and I heard his famous speech before the Congress after he was relieved of his command during the Korean War, I felt again his hands on mine and tears welled up in my eyes. The bitterness I felt in my own case and in his has not passed with the years.

The announcement of my retirement sparked a lot of mail and many words of praise in the press. I was offered several well-paying jobs in the aviation industry. The mayor and people in San Antonio asked me to come to live there in honored retirement. While the job offers were nice gestures, I couldn't accept them in good conscience. The service had been my whole life, and I had not ended it any blaze of glory. I was sick of the system that would allow a man to be vilified publicly, when his only crime was dedication to the cause of air power—something only a handful of people appreciated or understood in those days.

During my three-month leave, Mrs. Foulois and I decided to make our home near Atlantic City, New Jersey. We chose a frame house not far from the ocean and put a down payment on it. There I hoped to live in peace and forget Washington. However, I wanted to leave some kind of final word of warning as the outgoing Air Corps chief for the American people to ponder. I chose as my forum the *National Aeronautics* magazine and my theme was the inadequacy of our air strength to meet the growing danger in Europe and the Far East. I wrote:

> In playing the "numbers game" with the five other leading air powers, the United States is forced to take a back seat but without any of the privileges of the back-seat driver. If the Navy's airplanes are added to those of the Air Corps, and the total accepted as a representation of the air power of the United States, then we rank third.

I then pointed out that "it requires five years from the initial conception of the role of a military airplane until that airplane is ready to be procured in quantity." I reviewed the Air Corps needs for the next two years and warned that if we did not get the funds we needed, we would be reduced to one-half of our strength in first-line planes. As a clincher, I added:

> If the GHQ Air Force is to mean anything, it must be brought to full strength in modern equipment. As has been said many times before, an air force cannot be improvised after an emergency arrives, any more than a navy can be built after war begins. For that reason, the air force should, like the navy, be built up to and maintained at full war strength at all times. Unfortunately, the days of myths are past and we have no Cadmus among us to slay the dragon, sow his teeth and have fully-equipped fighting squadrons come sprouting. We must sow dollars instead. And we must not forget that after the dollars are sown it takes from a year to five years to reap any harvest.

The article was widely publicized and I received praise from people such as Sen. William Gibbs McAdoo, who had actually suggested I write the article, and condemnation from a highly vocal group of women pacifists. Although I had intended the article to be my final public plea before fading away into obscurity, I received a number of speaking requests from civic groups while I was on leave and decided to accept most of them. In an Atlantic City speech I blasted the pacifist movement and accused its followers of being ill-advised on national security matters although well meaning in their efforts to have peace in the world. I said that the activities of women's pacifist organizations were a greater menace to national security than the dissemination of Red propaganda and reminded my listeners that "military men don't want war but they know you can't have peace by abolishing our army or navy unless you abolish all armies and navies."

I was unaware of the moral support being given my views in Washington until Secretary Dern came forth with a strong statement saying that the Air Corps was far short of its reasonable requirements and urged the adoption of a five-year program under which a minimum of 800 fighting planes would be purchased annually. I learned later that General MacArthur had carefully studied the reports I had made during my four years as Air Corps chief and had "seen the light." He not only supported my analysis of the sad plight I predicted we would soon be in, but decided to argue the Air Corps case in my absence. He convinced Secretary Dem of the growing threat both in Europe and the Far East that I had forecast in the months and years before and showed that the intelligence reports coming in from Europe and Japan were more disquieting than ever. While the rest of the world belittled their efforts, those of us in uniform saw the potential for conquest developing slowly but surely.

General MacArthur's term as Army Chief of Staff was to have ended in November 1934. However, President Roosevelt had extended it for an indefinite period. In March 1935, MacArthur objected to the expanding powers of the Chief Executive over the size of the Army. He believed that the question of manpower numbers should be retained by the Congress, and he had said publicly that Roosevelt's attempt to capture this kind of control over the military was a dangerous precedent. As a result, the bill which would have authorized Presidential control over the Army's strength was killed in committee.

It has never been definitely established whether there was any connection between this conflict with the President resulting in the ending of MacArthur's tour as Chief of Staff, but he was subsequently shipped to the Philippines to what appeared to be a deliberately planned exile. I was sorry to see him go because I considered him the most brilliant and perceptive leader in this country. It appeared that he was headed for oblivion just as I was and for a similar reason. We had both bucked the political machines of our times and had come off second best. We had both done our utmost as military men to alert the Congress to the dangers of unpreparedness.

I felt a strange kinship with this man that I have never felt with any other superior officer during my service. Aloof, yet understanding; military to the core, yet warmly human. When you were alone with him, MacArthur was the kind of man you either deeply respected or hated with a passion. I not only respected him, I believed him to be possessed of almost godlike qualities. If ever there was a man of destiny, MacArthur was that man.

By December a new Chief of Staff, Malin Craig, was in command. He had been commandant of the War College and was no real friend of the Air Corps as MacArthur had been. When I returned to Washington at the end of my leave I tried to pay my respects but was unable to get an appointment although I received a nice letter from him regretting my retirement.

Since my retirement was to be effective on December 31, I decided to return to Bolling Field on Christmas Day to complete my flying time requirements. Snow was falling when I boarded the train at Atlantic City, and the weather was bad when I arrived at Bolling. I found Sergeant Sanborn, my crew chief, in his barracks and asked him to roll out-my O-38F. The sergeant was obliging, although I knew he thought it strange that the retiring chief of the Air Corps would go flying alone on Christmas Day in bad weather.

I took off and flew to Richmond under a sky that gradually cleared as if for my benefit. I refueled, and from there I followed the James River to Norfolk and turned south. I don't know why, but something was drawing me like a magnet toward Currituck Sound and the long finger of white sand that pointed southward. The visibility was perfect and the sky was bluer than I had ever seen it in twenty-eight years of flying. Below there were no cars on the roads, no trains chugging along the shiny rails, no smoke pouring out of the business district of Norfolk.

As I cruised along, my thoughts went back to those wonderful days at Fort Myer and San Antonio when every flight was a "first" and nearly every landing was a crash. I chuckled to myself over some of the flying scrapes I had gotten into and out of without a scratch. Flying machines had become more reliable now and certainly more comfortable to fly. Engines purred happily now instead of belching and coughing. We had instruments, radios, enclosed cockpits, landing lights, and warm flying clothes. We were talking of speeds of 500 miles an hour, altitudes above 50,000 feet, and nonstop distances of 8,000 miles. While there had always been those who said there were definite limits to how far, how fast, and how high man could go, it seemed as if there was always someone around to exceed those limits. But those of us who had flown since the beginning saw no limits. We thought man would go as high, fast, and far as he dared and that the only limitation was man's imagination. Someday someone would even go to the moon....

My daydreaming stopped suddenly. Where was I? I looked down at the glistening white mounds of sand broken by patches of dry brown grass waving in the wind. I was still headed down the coast. On the right I could see Albemarle Sound and the cluster of houses that was Elizabeth City. On the left, the sand spit stretched south-southeastward. I reached in my knee pocket for my map. Up to this point I had not used it but had just been flying by instinct rather than conscious direction. I spread out the map carefully and oriented myself with it. There was Cape Hatteras up ahead. Point Harbor and the quaint town of Mamie were where they were supposed to be on the right. On the left, the sand widened a little and then narrowed.

I had a strange feeling as I looked down at a flattened place in the dunes rising from the beach. There was the skeleton of an old shack there and I suddenly knew where I was. I began to circle. Below me was Kill Devil Hill, really only a mound of sand about 100 feet above the water level. Like a pilgrim going to Mecca, I had been drawn inexorably to that small deserted spot where aviation had been born thirty-two years before. “This is where it all began,” I said out loud to myself. “This is where it all began.”

I flew back to Bolling and landed after four and one-half hours in the air. Sergeant Sanborn was waiting for me. I didn't know it, but he had been worrying, too—so much that he had missed Christmas dinner at the mess hall. He signaled me to a parking space and I shut off the engine.

“Everything all right, sir?” he asked when the prop stopped.

“Yes, Sanborn,” I answered. “You can put the ship back in the hangar. I won't be flying it any more.”

The sergeant nodded and started to walk away.

“Ever,” I added, and tears welled up in my eyes.

The young sergeant understood. He helped me off with my parachute and pretended not to notice. “It's getting a little windy, sir. Going below freezing tonight.”

I grunted and reached for my handkerchief. Never again would I have the supreme privilege of flying a plane as a military pilot. Never again would I experience the ecstasy of solo flight where a man is truly the master of his own destiny, where he can be as near to his God as he can ever get during his lifetime. It was all over, and a sadness spread through me that was indescribable.

As I sat down to take off my flying boots, I felt a hard lump against my thigh. I reached down into a special pocket I had designed years before, closed my fingers around a steel object, and pulled it out. It was the horseshoe that Corp. Frank Scott had given me when I was recovering from my hernia operation in 1912. He had given it to me for good luck before he himself had been killed in a plane crash. It had certainly been good luck to me. I had never failed to take it with me when flying. While I never admitted to being a believer in good luck charms, I didn't want to take any chances. Call it skill, cunning, or luck, I had flown thousands of hours and survived. I was grateful, not only to Corporal Scott but to the hundreds of men like him who were the real backbone of our flying force—the dedicated enlisted men whose skills kept us in the air. They will never get enough praise. Without them there is no air force.

I sat alone in the pilot's locker room for a long time after changing clothes and then decided to call my office to see if anything was going on that I should know about. After all, I was still chief of the Air Corps and still responsible for whatever happened. Instead of a duty sergeant, Maj. Carl "Tooey" Spaatz answered the telephone.

"We're having a Christmas party, General Benny, at 5 P.M. in my home, and we'd be honored if you would join us," Tooey explained.

It was late afternoon but I decided to accept his invitation. When I arrived, almost my entire staff was there. They had drained a few bottles of Christmas cheer already and were in a high mood. It didn't take me long to catch up with them. We told flying stories until 1 A.M. when we ran out of liquor. I departed for the Union Station, caught an early-morning train for Atlantic City, and arrived home in time for breakfast with my wife.

Returning to Washington on December 31, I cleaned out my desk and shook hands with my staff. I expected Malin Craig or someone from the General Staff to drop by and wish me well but no one came. I thought I might get a wish-you-well note from Secretary Dern or one of his assistant secretaries but there were no messages, telephone calls, or letters for me. At five o'clock I signed out on the officer's register and left for Atlantic City by train where my wife waited for me. There had been no parades, no airplanes flying in review, no accolades from superiors, and no farewell dinners. My active military career was over.

It is not easy to change from uniform to mufti after thirty-seven years of service. I was fifty-six years old and healthy. It seemed to me that I could still be of service to someone or something, but a civilian job did not appeal to me. My heart was still in the sky and my soul belonged to the cause of air power. I decided that I would make myself available for speaking engagements on the subjects of aviation and military preparedness. My retirement pay was adequate for our needs, so I didn't have to go to work. My wife and I reduced our standard of living to fit my income, and I went on the road to preach the gospel of air power. During the next four years, I sought and accepted invitations to speak to civic, educational, and business groups in New Jersey, Delaware, Pennsylvania, and New York. There were about 150,000 people in those audiences but they seemed deaf to my pleas for an adequate air force. I recounted my experiences with Goering and Udet in Germany in the early twenties and warned that someday soon we would once more fight the Germans in the skies of Europe. I pleaded for grass-roots support of the Air Corps and warned that this country was ill prepared to match the growing Luftwaffe and the thundering German war machine that was building up.

It was discouraging to see the lack of concern on the faces in my audiences and then watch them scurry away after my talks to carry on "business as usual." But, beginning in September 1939, after Hitler's panzer units and Stukas invaded Poland and France, and Great Britain had responded by declaring war on Germany, the number of invitations suddenly increased. People began to listen to me and ask questions. My audiences were larger and they wanted to know what the government was doing about civil defense. When I replied "nothing," my words took on a new significance.

After September 1939, when my predictions came true, there were no more scoffers and disbelievers left. I was asked what should be done so often that I devised a plan in 1940 for civil defense for the Eastern seaboard that could be put into effect without government subsidy. I pointed out that there were no military installations of a defensive nature from Sandy Hook to Cape May and that the level topography and good roads of New Jersey made it a likely area for invasion. An enemy could make unopposed landings and head directly for the Philadelphia area where there were two Navy yards, an arsenal, and large shipbuilding yards in that industrial complex.

It wasn't long before civic and political leaders were so aroused by my speeches that things began to happen. The Republicans asked me to run for the House of Representatives from the Second District in 1941, and I accepted the nomination. However, my acceptance had some strings attached. "If I run for public office," I told the Republican Committee, "I will run on my record as a military man and will accept no financial support from the party. I do not want to be beholden to anyone for future favors." To my surprise, there was no hesitation on the part of the politicians, and I began a double-barreled campaign for both a civil-defense program and my own election to the Congress.

It was interesting, although maddening, to see what happens in a free election at the local level. My opponent was Elmer H. Wene, a Democrat who was running for a fourth term. He was a local chicken farmer who knew his way around the political barn. While I decided to base my campaign on state, national, and international issues, Wene based, his on *my* inexperience. "The war will not be won by aviators in Congress," he told his audiences, "where we need experienced statesmen, while we need all the airmen we can get in the air over the battlefield."

Wene's general plea for votes based on his six-year congressional record did not seem to impress even his own followers, so he must have decided that a smear campaign against me was his only hope. He subsidized some professional character assassins who researched every unfavorable newspaper story about me they could find dating back to World War I. Armed with old clippings and tired political clichés, he launched a vicious attack against me that included every kind of accusation from being a Fifth Columnist and a saboteur to having made myself a millionaire at the expense of the taxpayers while chief of the Air Corps. One of the curious accusations he concocted was concerned with a medal I had received from the King of Italy in August, 1934, during the visit of General Italo Balbo and his flying armada to this country: Wene accused me of having accepted it from

Mussolini himself and said that I had refused to return it to the Fascist dictator as others had done after war was declared. The inference was that I was a Fascist sympathizer and therefore dangerous to the nation's security. He was right in saying that I had not returned it and, now that he had brought up the subject, I would not have done so just on general principles.

I learned another lesson in the final days of the campaign before the election. I discovered that a congressman cannot be elected on the basis of national and international issues in this country. He is elected on the basis of local emotionalism and prejudice. While I tried to keep my campaign above the local level and avoid discussion of personalities, I was euchered into being put on the defensive by the vicious propaganda about my military service. The large issues of state and nation or the war were never faced or debated.

When Election Day came, I knew I was licked, but the vote count was close. I lost by only 550 votes in a largely Democratic district. I learned afterward that this narrow margin was almost precisely equal to the number of free chickens my opponent had given away the night before to the Negro voters on the north side of Atlantic City.

In the spring of 1940 President Franklin D. Roosevelt called for a 50,000 airplane program. In May 1940, I prepared a fifteen-page report on German air policies and German air plans for Gen. George C. Marshall, the Army Chief of Staff, and offered some suggestions with respect to the mass production of aircraft. Marshall replied that "confidentially, as a result of your suggestions, I am having studied the development of a multipurpose plane of the air cruiser type capable of performing both combat (pursuit) and bombardment missions," and added that he was also "looking into the matter of relaxing present close manufacturing tolerances in order to facilitate mass production."

I never found out what actually developed as a result of my voluntary study, but in 1942 I was asked if I wanted to return to active duty in Washington. I declined because they had a nonproductive desk job in mind. Besides, I interpreted the offer as only a kind gesture to an old man of sixty-three who had been a martyr in the fight for air power and should now be rewarded for being right. There still seemed to be a guilty conscience among the old-timers after the death of Billy Mitchell in 1936, and they now remembered many of the predictions I had also made that had suddenly come true.

I felt that I could be of much more service to the civil-defense effort and devoted my entire time during the war to running the program in New Jersey. I helped Fiorello La Guardia, the mayor of New York, in his civil-defense program and also helped establish the Civil Air Patrol which did such a magnificent job of sub patrol in the early days of the war.

When World War II was over and the world knew that it had been won by a unified military effort that included the largest, most powerful air force in history, I wondered what would happen next. Would we let our military organization disintegrate as we had done after World War I? Would we let the machines of war be turned into junk and cut our military strength back to a skeleton force? Would we forget the perils of communism and be lured into thinking that our Johnny-come-lately allies, the Russians, would give up their own plans of world conquest?

The world knows the answers now. The military services were cut so badly they could not even muster enough men to put on a decent parade anywhere. The Air Force became a separate service in 1947 but immediately found that it had to struggle for funds in competition with the Army and Navy. And then came the Berlin air lift, where the Air Force proved it could keep a city alive with air power. The "police action" in Korea followed and the Western nations of the world aligned themselves against the Communists in the Far East. Again, after hostilities ceased, we tore down our forces to dangerous levels.

Now we are deeply involved in another conflict in Southeast Asia, and it seems that as I read the daily reports coming in from Vietnam my life has gone full circle. We are fighting a holy war in the Oriental jungles again just as we were doing in the southern Philippines against the Moslem Moros in the early 1900s. The enemy still owns those jungles and is better prepared than ever before with weapons. This time he has a strong sustaining ideology. He also has huge reservoirs of manpower behind him and the patience of centuries to pit against the traditional impatience of the Occidental. It seems to me that the dominant force that stands between us and a long-drawn-out war is our air power. If we let ourselves be dragged further into jungle warfare on the ground, we must prepare for at least a decade of armed struggle in which thousands of American boys will fight and die. Our strategy must be changed to allow the airman more freedom to operate and exploit the machine that knows no geographical limitations.

As I look back over the more than six decades since the airplane was born, I see that the aeronautical sciences have come a long way. The airplane is now complemented by missiles and space systems. Since I have lived through the entire life history of aviation and was a participant in many of its earliest successes as well as its failures, I cannot resist taking a look backward occasionally.

I am first impressed with the fact that at each step along the way many people thought that we had reached the pinnacle of our technical capabilities. By World War II we had run out of superlatives. We had the greatest air fleet in history, we had established air power as a dominant element in modern warfare, and even the most optimistic citizens were saying that small progress would henceforth be made in aviation and that the great strides were behind us. How wrong they were. First came the jet era, then the ICBM era. These were followed by Sputnik and our magnificent space program. Notwithstanding the experience of these, however, there were people who said that certainly now we had reached a plateau and that rockets would supplant aircraft, which had reached, if not exceeded, their growth limits from a technical and useful operational point of view. By 1950 some were advocating that the Air Force would move entirely into space or missilry. However, a few bold, imaginative thinkers in blue uniforms recognized that the airplane was no more on a plateau in the early 1950s than it had been in the 1930s or in the 1940s.

The Air Force, led by those who foresaw the new frontiers that missile and space technology had opened up for aerospace power, took a new look at the beginning of the 1960s to see what might be done for the airplane with those technologies. The results were startling. Not only had we not reached any limits to our technical or operational possibilities, but we foresaw tremendous strides in aeronautics as well as in space and missiles. If anything, the growth of the missile tended to level off but not that of aircraft.

As I write these words, the Air Force is trying to get back on the growth curve in so far as aeronautical systems are concerned. We can see coming rapidly ahead of us all types of vertical-lift flying machines, with the tremendous military and economic advantages that these promise. Flight in the atmosphere, at much greater speeds than are now commonplace, will be a reality within the next decade. In our space and missile effort we have already established velocities well above Mach 20 and have penetrated heat barriers that are associated with these tremendous speeds.

In 1946 Dr. Theodore von Karman gave the Air Force its start into the space and jet age with his famous study *New Horizons* in which he forecast the future for many years to come. This was the same Von Karman I had known in Germany in the early twenties and whom I had advised even then to settle in the United States, so that he could give vent to his tremendous imagination. His forecast proved unerringly accurate through the fifties and sixties. Today most Air Force leaders can clearly see a similar trend with further new horizons that are just as dramatic as those we crossed during the last fifty years. Anyone who says that air power is being overtaken by missile or space power is either naive or unduly prejudiced. He certainly doesn't know the facts of modem technology or understand the contribution that transports traveling at 5,000 to 6,000 miles per hour, or metroplanes able to service the great metropolises of the Eastern seaboard with vertical takeoffs and landings can make to both the military and the Great Society.

The prospects of new horizons in aeronautics are only outmatched by those that the space adventure has opened up before us. As I look at our space achievements, I marvel at the technology that permitted us to achieve these feats as rapidly as we have, but I do not find anything fundamentally new in the impact that these technologies will have on the classic principles of war or on the role of national power during peace. It is only a question of degree and how these are interpreted in terms of the greater access provided by the new media.

When one has lived as long as I have, one tends to look at things through the long eyes of history and dwell on the fundamental issues rather than focus on the current problems of today or those anticipated for tomorrow. The visionaries of space are no different from the visionaries of my day so far as their theory and their objectives are concerned; their purposes are equally valid and their views equally true. Space will provide the new dominant media in the seventies and eighties for both peace and war that air provided in the forties and the ballistic missiles provided in the fifties and sixties. Why do I say this? It is all really very simple.

History has shown that strategy, tactics, concepts, and even politics, as well as world-power positions, all eventually adjust to technology. The key advantage that space offers from the airman's point of view is a new order of magnitude in accessibility. In the early days before World War II the navies and fleets of the world with their then modern battleships were just as impressive as spaceships are today from a technical point of view. They commanded the seas and hereby had the greatest access to the shore and to ports, either for commercial purposes or to deliver military forces. Those nations that could deny access through better naval power controlled the seas, and this factor established not only the role of naval power in national security but also the relative power among nations on the basis of who possessed the best naval forces.

In later years the airplane took over from the battleship as the dominant weapon system. Air power came into its own in World War II after struggling through World War I and between the wars. Technology conferred on the air forces the decisive role of greater access. Air forces not only enjoyed access to the seas and the shore, but also to the hinterland. They could deliver bombs, leaflets, cargoes, or personnel anywhere on earth without regard to geographical barriers. Granted our ranges were limited and our possibilities poor at first, but these were later overcome, and when that happened air power defined the relative power position of all nations.

The ICBM improved access by a new order of magnitude. We had global access with aircraft, but the time and the ability to penetrate an enemy's hinterland limited its exploitation. With the advent of the ICBM the enemy's capability to deny access to his vital centers was again reduced to almost zero, where it has rested until today, and will remain until new technology provides us with an effective antiballistic missile system. The spiral, however, goes on from year to year. First the offense, then the defense, gets a slight lead over each other, and finally the game moves to another arena and starts allover again. It seems inevitable to me that this is what will happen in space.

Space has already followed in the footsteps of World War I aviation in that it first furnished us access in terms of reconnaissance. We were able for the first time to see targets in the hinterland of Russia that we could reach but did not know where they were. The same thing happened with the World War I airplane. First came the reconnaissance phase, then ultimately the air-to-air phase, and finally the air-to-ground or bomber phase. I forecast that the same evolution will occur in the space media. We have already witnessed the dominant role of reconnaissance satellites. The next logical evolution in military space systems will be to attempt to deny one another's satellites by space-to-space military action and finally someone will devise a weapon system or a method of applying power or constraints from space, which is more effective than by missile. When this happens, the full cycle that the aircraft ran will have been accomplished in space and space weapons will become the dominant military system.

There are three great factors in air power, and, for that matter, in military and national power. They are those weapon systems that technology provides the country at any time or in any era, the concepts, tactics, and doctrines by which they are used, and the organizational principles under which these weapons are developed and exploited in the field. It is these last two items which are most neglected and least understood.

For more than sixty years I have seen the services argue among themselves and take varied positions on strategies and concepts. Almost invariably when these issues become heated one discovers that the position taken by the proponents just happens to be that which their hardware is best equipped to carry out. When all the arguments are over and the results are in, or when we are forced actually to carry out in war the concepts that we advocate, we invariably come back to the fundamental principles laid down by such great military writers as Von Clausewitz, Napoleon, Frederick the Great, Douhet, Mahan, and others. As I have said, changes in technology cause these principles to vary in their application or interpretation, but not in their fundamental tenets.

One of the basic principles of warfare, applicable to land, sea, or air forces, and which will be applicable to space forces, is that so long as there are wars, the primary objective of the military organization must be to find, defeat, and destroy the enemy's forces. This simple statement has been referred to in many different ways and most recently is called "damage limiting." Whatever the semantics, the facts remain the same. It is the military mission in so far as possible to destroy the military might of the enemy in war. Once this has been done, the enemy nation, being powerless either to deny you access for whatever your purposes or to achieve access to your vital areas or your forces, must bow to the will of your political authorities. But until the armed forces are destroyed, the enemy nation will always be in a position to negotiate or even win.

I mention the above time-honored principle because I have been disturbed in recent years over the tendency of the military to place deterrence of war ahead of war waging and war fighting as their main objectives. This is fallacious and suicidal reasoning in my view. The Air Force and the military as a whole must remember that deterrence is a perfectly plausible and valid national and political objective, but it cannot be the primary Air Force or military objective. The purpose of military force is to protect the country. if deterrence fails, and as such the primary Air Force objective must be to fight and have the capability to counter enemy forces in combat. Deterrence is a by-product of a valid and credible counterforce capability. The converse, however, is not true for it is possible to deter a military confrontation without necessarily having a counterforce capability. If the enemy is not deterred, however, and we have only a minimal deterrent force to fall back on, we would have no flexible response forces and would have to surrender to the enemy—an unthinkable but likely possibility.

In the early days of World War I and II the older services and political echelons wanted air power to concentrate on delivering weapons on the enemy ground forces. I did not agree with this because I believed that the first thing we had to do was to deal with the enemy air forces and that the initial phase would be, as it was in the Battle of Britain, a battle between air forces for air superiority. When this battle had been won, only then could we begin to worry about engaging other types of targets.

I see the same mistakes being made again in space. Modern-day Billy Mitchells want to leap into space vehicles and immediately use these to deliver hardware against the enemy. They fail to recognize that if history repeats itself, as it always has, the initial phase of space warfare will be in essence a battle for space superiority—for control of the space media, at least in those areas where it can be profitably utilized and/or be dangerous to us. I believe we should concentrate far more on developing a space-to-space capability, an anti-satellite capability—and on inspection and destruction of space vehicles, and let the generations ahead of us worry about utilizing these vehicles for offensive purposes once we have achieved the means of controlling space.

I believe that air power must take the lead in innovations. Our Air Force must innovate not only in response to innovation by others, but also in order to force others to innovate in response to us. It is only when our nation takes the positive initiative in this eternal struggle for world power that one can remain and retain for the United States technological superiority and, through that, a dominant, war-winning, deterrent posture as we have in the past.

Since I am the last of the first military airmen in the United States, I enjoy a sort of special status in the military aviation world. I am asked many questions about the Air Force of today and tomorrow and what I think we should do to win the Cold War. While I would like to answer by referring to the past, I find that too many people know little or nothing of what happened before they entered the service—the struggles, the frustrations, the heartbreaks of yesterday's airmen. So I try to appeal to their knowledge of today's events and place the East-West struggle in their perspective instead of mine.

As I look on our world today, I see a great decline in the ability of the United States to control or influence world events despite our formidable industrial capacity and our present military superiority. This is owing, in part, to the recovery of European nations since World War II, the rising current of nationalism, and an increasing self-confidence leading to the pursuit of independent policies by more nations. These things have greatly affected our national planning in general and our Air Force planning in particular.

But there are other complicating factors. Nations formerly docile and content to live under Stone Age conditions have been awakened and are exerting their personalities on the world scene. Limited wars have erupted and what would have been local disputes two decades ago now become international disputes involving coalitions of nations such as the United Nations, Organization of American States and the North Atlantic Treaty Organization. Ironically, advanced technology does not enable the world powers to solve these disputes any faster or any more permanently.

It is certainly no secret that the United States Air Force is a vital part of our over-all national strategy. But from my observation point on the side lines, our Air Force is declining at an alarming rate and may soon be as weak in the late sixties as it was in the early thirties when we were trying to build bombers with public-works money. The Air Force may soon, if it does not already, have fewer flying machines than the Army and the Navy. Every day our bomber force dwindles into obsolescence and no new bombers are coming to take their place. We are losing fighters over North Vietnam at an alarming rate and our air-defense fighters at home are gradually fading away. Thus, we are losing the flexibility we need to fight limited wars at the most critical time in our history.

I consider this nation to have reached that point in its military aviation posture where we are no longer fully protected by the shield of American air power. Something must be done and done quickly before we are involved in World War III and find ourselves in the same sad shape we were in at the beginning of the two previous world wars.

I am greatly alarmed at the rapid decline in the number of strategic and tactical manned aircraft in the Air Force inventory. While intercontinental missiles may be valuable in an all-out nuclear war we have never fired any in anger and we will never be sure that they can hit their assigned targets until they are let loose.

We live in an age of undeclared non-nuclear limited wars. Missile systems with nuclear war heads cannot be used. The only air weapon of any value then is the manned aircraft because it can give us the shades of military power required in counterinsurgency or limited wars.

It would be nice if a single plane could be built that would have all the characteristics of a bomber, fighter, and reconnaissance plane. It is my opinion, however, that such a plane cannot be built unless there is some compromise in the performance of each of these missions. The civilian powers-that-be in the Pentagon say the new F-111, with its ability to fly fast and slow, high and low, will be the answer. I don't believe it, and it is my feeling that this calculated risk being taken by the computer-minded young scientists running the Defense Department today has placed this country in the most serious military position it has been in since the Revolution. If they are wrong, and I think they are, their reliance on a single multipurpose plane would be like having the Navy depend entirely on submarines or the Army on armed helicopters. If they are right, and the F-111 proves out in extensive testing, these planes still won't be in the hands of the fighting squadrons in any quantity until the beginning of the next decade.

I say the F-111 is not enough. We need an advanced manned bomber capable of flying long distances without in-flight refueling, able to operate at extremely high and low altitudes at supersonic speeds, and be able to carry either nuclear or conventional weapons without extensive modification. We also need new manned interceptors able to scramble rapidly and knock out enemy bombers far enough from our borders to eliminate danger from air-launched missiles. Such an interceptor will be needed as long as any potential enemy can attack us with his own manned aircraft. The F-12, which could be the answer to this requirement, is not being produced in quantity because the computers tell the chair-borne civilian scientists we can wait a while.

We must also improve the air-lift capability of our Armed Forces. When the mammoth C-5A comes into the inventory soon, we will have gone a long way but, in the meantime, the logistics system of the Armed Forces must be able to support large ground and air units in several trouble spots in the world at the same time. The war in Vietnam has strained our air-lift resources to the utmost. If there is another Congo operation, more trouble in India, and a new crisis in the Caribbean, our air-lift capability will be dangerously deteriorated. The new C-141s now in Air Force hands are arriving literally in the nick of time. The C-5As may be too few and too late.

Another Air Force requirement is for a supersonic reconnaissance plane with great range and altitude, able to penetrate enemy defenses, and capable of relaying their information "live" to decision makers or to deliver processed film. As long as there remains a missile threat, there will be a need for a high order of continuing strategic reconnaissance, whether by manned aircraft or manned spacecraft.

While I am a strong advocate of the manned aircraft, I believe work must continue to improve our missile force. A more sophisticated missile that will carry a larger-yield war head and improved penetration aids should be developed to supplement and eventually replace the Minuteman force. We must consider a completely mobile missile force consisting of land, sea, and air launches. Nuclear rocket propulsion with virtually unlimited range capabilities must be developed. And the tactical missiles must also be improved to contain electro-optical guidance systems. All of these things are desperately wanted by the Air Force. Very little is being done to get them because military professionals are unable to convince their short-time civilian bosses that the threat is great, time is short, and the need is becoming more desperate every day—all items that can't be cranked into the brainless computers. No computer has been or will be devised that can accurately forecast United States military needs. Since we are not an aggressor nation, no attack plans are made; our plans and tactics must be modified and adapted to meet the situations thrust upon us by nonmilitary policies.

The continuing and significant progress in space technology has increased military awareness that competence in space is essential to national security just as it is to national growth and prestige. Soviet achievements and the fact that they have surprised us so many times since the first Sputnik present the possibility that the U.S.S.R. will achieve a significant military space capability that we cannot match. Soviet "peaceful" efforts may one day be accompanied by hostile demonstrations in space that will give them the ability to control space and rule it. If the United States does not have the capability to negate other space vehicles, we will be at a distinct military disadvantage, in spite of our earth-based strategic deterrent strength.

The Manned Orbiting Laboratory (MOL) program of the Air Force is, in my opinion, the key to future military space operations. It will provide answers to many of the unknowns about man's ability in space—answers from a defense viewpoint. I think it is possible that space systems may prove to be the most economical and effective way to perform several basic military missions. For example, a laser-beam weapon operated by a man might be used with great precision, like a surgeon's knife, to cut out enemy military installations with little damage to nearby towns. The space pilot, on routine patrol, will be able to inspect a potentially dangerous enemy satellite, neutralize it, and report the threat to earth. He will be able to spot future trouble as it begins to develop on the ground, such as unusual deployments of ships, submarines, planes, missiles, or ground forces. This kind of information, received early enough, could help ward off international crises such as the Cuban crisis of 1962.

I am confident that by the end of this century technology will have advanced to the point where space operations will be carried on as we do air operations today. The aerospace plane of tomorrow will be able to take off from a conventional airport, zoom into orbit, and return at will. It will be used to resupply orbiting space stations, for space rescue, or any other kind of mission that conventional aircraft now perform. The only question is will that aerospace plane be ours or belong to someone else?

The ever-increasing complexity of military equipment reminds us all of the need for properly qualified people to run it. In a military environment requiring imagination, ingenuity, and education, the future requires people who do not fear automation or machines but who will have the dedication and the will to build better devices and master them. The same imagination and ingenuity applied to the development of military weapon systems must be applied to the management of men and women. How best to use men and women in conjunction with computers, how to train, motivate, and retrain them to meet the new and greater demands for the future are questions that must be answered. For it is really the men and women who hold the key to the Air Force of tomorrow—not the machines.

In 1959 I returned to the Air Force after twenty-four years of what I call "civil confinement." In October of that year my wife became seriously ill and passed away a short time later. At that time Gen. Thomas D. White, then Air Force Chief of Staff, invited me to live at Andrews Air Force Base near Washington, D.C., and spend whatever remaining years God gave me to preach the gospel of air power. I accepted his invitation and have traveled almost a million miles in this country and overseas since then. Naturally, my talks include references to my experiences in the old days, but I always relate them to our world troubles today and encourage my audiences to look ahead at where our great nation is heading.

I am asked many times for my views about our involvement in Southeast Asia, and I have some opinions that are sometimes completely opposed to press reports regarding air policies and plans of our present administration. As the elder member of the Air Force congregation, I am encouraged to say my piece because it is the sermon that my active-duty brothers want to hear but are not free to support under the present muzzling rules of the Defense Department. Since I have absolutely no military future, I do not feel any obligation to keep silent. As a senior citizen I feel compelled to point out the dangers our beloved nation faces.

We are living in an age in America now where we have more military experts in the press than in uniform. But while we debate the basic issues in the open forum as we are privileged to do in our society, American lives are being lost again in Southeast Asia. Our present publicized political policy requires us to fight a polite, constrained war in the jungles against a relatively primitive but highly motivated enemy. For the past ten years our Armed Forces have been required to fight with one hand tied behind their backs, so to speak, and it is about time they were allowed to do their fighting without this political handicap!

Our greatest publicized fear seems to be that the Chinese Communists will join the North Vietnamese. I don't think they will. Why not? Because they fear American air power which would be surely unleashed against them.

What is the solution to the Vietnam dilemma? Strategic bombing of selected military targets in North Vietnam, mining of the harbors, and an air-sea blockade that will absolutely prohibit the movement of any kind of war goods into the country and cut off their communications with the world. We should announce the blockade and then put it into effect. We should expect sharp criticism, but stand firm in spite of it. The British will condemn us but the Canadians, New Zealanders, and Australians will back us. The South Koreans and Filipinos, recognizing that we are fighting their war, will give us all the help they can. The rest of the Allies will shake their heads and wring their hands. In spite of this, we should, and we must, take a firm stand. Call it "interference" or "escalation" if you want, but it's the only way to win in Asia. If we continue to fight a polite war, to defer and appease, we are defying the lessons of history. No wars have ever been won by any nation that followed a defeatist course of action such as we are following now. The Communists are testing our competence and our will to win. They are testing our ability and willingness to use correctly the great military power we have. Most important, they are testing our national courage.

Besides the weapons of war and the permission to use them, this nation needs a strengthening of its moral, social, educational, and physical fibers. We should return to a way of national thinking where it is popular to put the flag out on national holidays, sing the national anthem, say our prayers, and respect our elders. We must clean up corruption, crime, and prejudice at home before we can expect the rest of the world to accept our ideals.

What is the answer to our present-day moral, social, educational, and physical problems when analyzed and evaluated against corruption, crime, and Communist infiltration?

The answer, in my belief, is universal military service—the type of military service basically advocated by Robert S. McNamara, Secretary of Defense, in his address before the American Society of Newspaper Editors on May 18, 1966; the type of military service basically applied by the Armed Forces in the Civilian Conservation Corps (CCC) established by Franklin D. Roosevelt prior to World War II, primarily to care for so-called juvenile delinquents—a characterization, incidentally, too freely applied to all our young people.

The CCC was established to care for, educate, and assist our fine young men whose families were in dire financial straits. I, myself, at the age of eighteen, would have been among this group in that I could have been characterized as a so-called juvenile delinquent if running away from home, joining the Army, and being shipped off to the Spanish-American War are the qualifications. But looking back at the training, the discipline, the moral guidance, the education, and the opportunities which the military service made possible for me—to that, and to this wonderful country, I give my everlasting thanks and gratitude. All of this has fostered my deep concern, genuine interest, and high regard and affection for our young people, and I would advocate for them only those things which would contribute to their welfare, happiness, and security.

Having had a number of CCC camps under my jurisdiction when I was chief of the Army Air Corps, I want to say here that thousands of these so-called juvenile delinquents served in World War II with brilliant military records, both as enlisted men and as commissioned officers. Many of them continued to establish brilliant records in the Armed Forces after World War II, and those who returned to civil life became outstanding leaders in their local communities.

Perhaps it is wrong for military men my age to attempt to give advice to my country; but if I don't, it is a certainty that no one on active duty will do it. Our men and women in uniform, the men and women from whom military victory is expected, are both handcuffed and muzzled today. They cannot exploit their weapons to the fullest, and cannot speak out if their thoughts are in opposition to those in political and diplomatic power. They are discouraged from writing and cannot lend their voices to the forum of public opinion. This is an unwise, dangerous trend that truly places us on the threshold of national disaster.

My lifetime has spanned the entire history of powered flight. I have seen the airplane grow from a primitive bundle of wire, fabric, and wood to an instrument of national power. I am proud to have been a part of the organization that has done more to bring it to its present state of development than any other—the United States Air Force. I am pleased that many of the developments I encouraged during my tenure in uniform contributed to our present success in flight. But of all the things I have seen and done, the one achievement I want to be remembered for is the establishment of the "can-do" spirit that has become traditional among our American airmen. It is that spirit that we need on a national level to accomplish the world peace we so earnestly seek.

It is a national objective that an American will land on the moon by the end of this decade. I think we will accomplish that objective. When we do, I will have the supreme satisfaction of knowing that the battles I fought for American air power were not fought in vain. I will think that I had a small part in giving that lucky astronaut a boost on his journey to the stars.

Epilogue

The activities of Major General Benjamin D. Foulois were not front page news after he retired as Chief of the U.S. Army Air Corps in 1935. He was known as one of the first military pilots who had achieved some notable aviation "firsts" but had also been criticized severely for being in charge during the "air-mail fiasco" of 1934. He moved to New Jersey and ran unsuccessfully for Congress. When his wife became seriously ill in 1958, she was admitted to the hospital at Andrews Air Force Base in Maryland and he moved to officers' quarters there to be with her. She died in 1961.

When he recovered from his grief, General Foulois, an avid reader, became increasingly concerned about world events and the serious Communist threat—so concerned that he volunteered to be a public speaker and "preach the gospel of air power." Those in charge in the Air Force soon began to realize that this Grand Old Man of Military Aviation was not only still alive but very vigorous and capable of making significant observations about military affairs which were not held back by fear of censure by the Department of Defense or concern for his own personal future. He received and accepted scores of invitations to speak at events honoring aviation pioneers or commemorating aeronautical achievements.

Gifted with a superb memory for names, dates and details of events, General Foulois traveled thousands of miles during his seventies and early eighties to speak to scores of attentive listeners, always with a touch of humor, about his early flying experiences. He left his audiences thoroughly impressed with the problems that he and his early contemporaries had met and solved. But he always had a deeper message for his audiences than to remind them of their aviation heritage. He warned them about the dangers of a complacent national mindset about the growing Communist threat of those days and of the nation resting on the laurels of past military victories.

Many writers, after hearing General Foulois speak, asked if he would collaborate with them to tell his colorful life story. He turned them all down and said he was too busy. He said he "had the constitution of a billygoat" and had about 20 years more to go before he would slow down and write a book.

At the behest of John W. King, then governor of New Hampshire and his sister Mary King Reeves in early 1965, he finally relented. He contacted the McGraw-Hill Book Co. and announced that he was ready to work with a writer who met three specifications: be an Air Force officer on active duty, an active military pilot, and have at least one book on aviation published previously.

It was my honor to be selected to assist this great man whose life had spanned the entire history of powered flight from Kitty Hawk to Cape Kennedy. From the moment we began work, my respect grew for his wisdom and foresight, his knowledge of national and world affairs, and his personal philosophy about life. He had been an excellent record-keeper and his files contained vital proof of his claims about people and events if anyone were to doubt.

We finished our work together in July 1966. The last time I saw him was in September of that year in the hospital where he had gone to get what he called his annual flight physical exam. He suffered a heart attack there that left him speechless and paralyzed; however, his mind was active and he tried desperately to form words of greeting. The only sign of recognition or an answer to a yes/no question he could give me was a hand squeeze—one squeeze for yes, two for no.

On April 25, 1967, the stout heart of this tough little fighter of 87 years ceased forever. He had finally gone to join his flying comrades whom he had outlived to earn the title of "the last of the first." The United States Air Force had lost a leader who had accomplished more to make it a power for peace than he had ever been given credit for publicly. I lost a friend.

Although he did not see the book in published form, he did

see and approve the entire manuscript before he was hospitalized. He had many professional disappointments in his lifetime. He suffered greatly from criticism leveled at him by superiors who neither understood the problems of the airman nor the potential of the airplane. He had failed to find Pancho Villa in Mexico in 1916; had been maligned by a subordinate named Billy Mitchell during World War I, was blamed by President Roosevelt for the deaths of air mail pilots in 1934 and was forgotten by the men he had led in the days between world wars.

But General Foulois left us a priceless legacy. He had laid the groundwork for American air power which contributed much to winning World War II. He made each disappointment during his career into an ultimate victory by calling attention to the reasons for his failures. He had failed to find Villa because the airplanes would not climb high enough; there were no American planes in European skies during World War I because the government had not planned ahead for war in the air; pilots crashed during the air mail days because the nation had not provided planes equipped with lights, instruments or radios. After each failure, he pointed with alarm at America's lack of preparedness in the air but to little avail. To his eternal credit, he never gave up, hoping that someday his views would be accepted.

Chafing at public apathy, the Little General thought of other ways to accomplish what he believed were necessary for the public defense. He encouraged clandestine research by his pilots; he persuaded the aircraft industry to spend company funds to build prototype planes without promise of a government contract; he cajoled Congressmen into voting for flight pay for aircrews and decent housing for his men. When he left active duty in December 1935, he left behind proposals for almost every plane that saw combat during World War II and the basis for a sound organization that could be expanded a thousand fold when the need arose. The young officers he led and inspired became the air generals who masterminded the eventual victories of World War II.

Americans owe a debt to Major General Benjamin D. Foulois it can never repay. These memoirs will be a constant reminder of that debt.

Carroll V. Glines
Colonel, USAF (Ret)

Acknowledgments

No book becomes a reality without a team of dedicated and talented people. I would like to thank those that helped bring this important book to life. Each of them shared the dream and desire to see the amazing story of Benjamin D. Foulois come to national awareness as much as I. My sincere thanks to you all.

Alan Ericksen
Forté Public Relations, Inc.
Laura E. Hall
Roger G. Miller, Ph.D.
Mandy Mulliez
Bill Pemberton
C. Scott Sykes
Hisham Syed
Fran Ussery
Beth Wilbins
Robert Wood

Carroll V. Glines
February 1, 2010
Dallas, Texas